Materials Forming, Machining and Tribology

Series editor
J. Paulo Davim, Aveiro, Portugal

More information about this series at http://www.springer.com/series/11181

Dhorali Gnanasekaran • Venkata Prasad Chavidi

Vegetable Oil based Bio-lubricants and Transformer Fluids

Applications in Power Plants

Springer

Dhorali Gnanasekaran
Dielectric Materials Division
Central Power Research Institute
Bengaluru, Karnataka, India

Venkata Prasad Chavidi
Dielectric Materials Division
Central Power Research Institute
Bengaluru, Karnataka, India

ISSN 2195-0911 ISSN 2195-092X (electronic)
Materials Forming, Machining and Tribology
ISBN 978-981-10-4869-2 ISBN 978-981-10-4870-8 (eBook)
DOI 10.1007/978-981-10-4870-8

Library of Congress Control Number: 2017945271

Printed on acid-free paper

This Springer imprint is published by Springer Nature
The registered company is Springer Nature Singapore Pte Ltd.
The registered company address is: 152 Beach Road, #21-01/04 Gateway East, Singapore 189721, Singapore

Preface

One of the goals of this book is to provide information on the environmental benefits and the importance of vegetable fluids as biolubricants and of bio-insulating fluids in power plants. Vegetable oil naturally is an excellent biodegradable, non-toxic, and renewable lubricant and insulating oil, owing to its functionality and structural arrangements. In a power plant, its use as a transformer fluid and lubricant can play a vital role in reducing negative impacts on the environment.

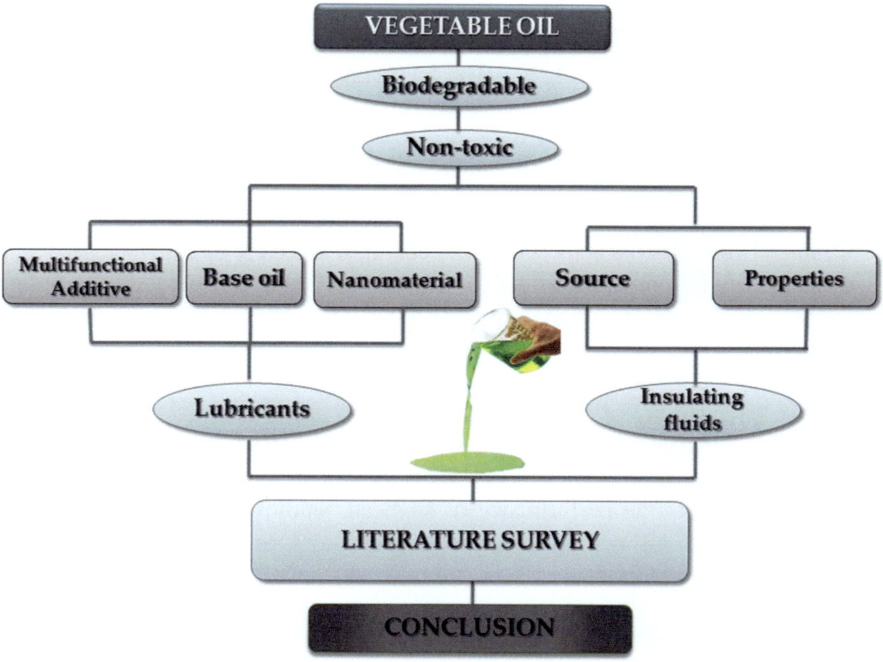

Fig. 1 Main goal of the book

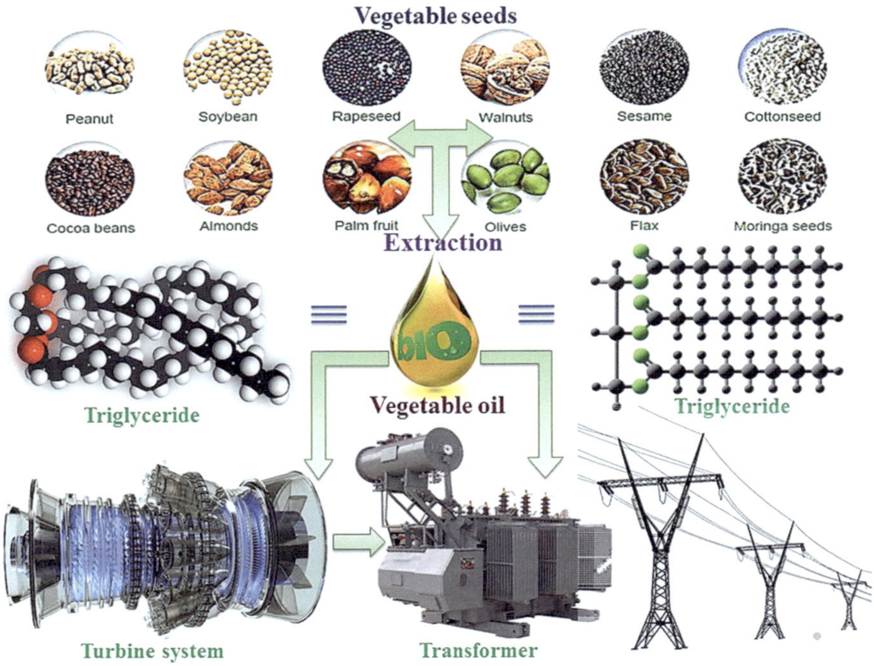

Fig. 2 Scientific approaches of the book

The book is divided into three parts: Part I comprises Chap. 1, which explains green fluids from vegetable oils in power plants; Part II consists of Chaps. 2, 3, 4, and 5, which provide the cumulative statistics of vegetable oil for lubricating applications; and Part III consists of Chaps. 6 and 7, which deal with vegetable oil as a liquid insulator in the power sector. Figures 1 and 2 give the blueprint of the book in brief.

- Chapter 1 provides a conspectus of the chemical structure of vegetable oils and its suitability as lubricants, as well as insulating oil in power plants. In addition, the pattern of lubricant usage in India and statistical developments of biodegradable oil in the last two decades are discussed.
- Chapter 2 describes the terminology and significance of biodegradation, toxicity, and the renewability of vegetable oil, especially the elucidated status and scenario of Indian use of lubricants, and concurrently, the tabulated chemical, physical, and biological properties of the oil.
- Chapter 3 covers the performance of soybean, sunflower, jojoba, and natural garlic oils, palm oil methyl ester, lipoate esters, and rapeseed oil as multifunctional additives in lubricants.
- Chapter 4 presents the improved properties (pour point, viscosity index, EP) of polymers (oleats, diisodecyl adipate, polyalphaolefin, polymethyl methacrylate, ethylene vinyl acetate, ethyl cellulose, styrene-butadiene-styrene, polysulfide, and polyester) as additives in lubricants.

- Chapter 5 discusses the tribology properties of friction, wear behavior, and worn surface analysis of CuO, ZnO, boron nitride, WS_2, and TiO_2 and graphene as additives in lubricants.
- Chapter 6 describes the chemistry of natural esters and their extraction process, based on various vegetable seeds. It also describes the use of natural esters as liquid insulators in transformers, and explains the current research scenario on vegetable oil in the power sector.
- Chapter 7 includes various properties (physical, chemical, and electrical), antioxidant additives for vegetable oil properties, performance and suitability of vegetable oils as transformer oils, challenges, and technical difficulties.

Bengaluru, Karnataka, India

Dhorali Gnanasekaran
Venkata Prasad Chavidi

Acknowledgments

The authors are grateful to the employees of the Central Power Research Institute, Bengaluru, India, for all their support and encouragement at the various stages of the preparation of this book.

Contents

About the Authors

Dhorali Gnanasekaran, M.Sc, M.Phil., Ph.D. is currently a scientific officer at the Dielectric Materials Division, Central Power Research Institute, Bengaluru, India. He received his Ph.D. in polymer chemistry at the University of Madras under the supervision of Dr. B.S.R. Reddy in the CSIR–Central Leather Research Institute, Chennai, India. After completing his doctorate, he joined the University of Pretoria, South Africa, for postdoctoral studies in the prestigious South African Vice-Chancellor's Postdoctoral Fellowship Program in 2012. His research interests include the preparation of eco-friendly polymer nano-composites for gas permeation studies and biodegradable polymers and additives as eco-friendly/green lubricants/insulating oil for electric power generation. He has published 20 research articles in peer-reviewed international/national journals, one review article, six book chapters, and one book, and he has been a presenter at 10 national/international conferences.

Venkata Prasad Chavidi, M.Sc., Ph.D., is a scientific officer at the Dielectric Materials Division, Central Power Research Institute, Bengaluru, India. He holds M.Sc. (2005) and Ph.D. (2010) degrees in Polymer Science and Technology from the Sri Krishnadevaraya University, Anantapuramu, Andhra Pradesh, India. He has worked as a research professor at the Changwon National University and Korea University, South Korea (2011-2013), and was awarded a prestigious South Korean Brain Korea 21 Fellowship (BK21). His current main research areas are solid and liquid dielectric materials (mineral oils and vegetable esters), and his other research interests are polymer matrices for pervaporation drug delivery and fuel cell applications, polymer blends and natural fiber, and polymer composites. He has published some 40 research articles in national/international journals and has presented at some 20 national/international conferences.

Part I
Introduction

Chapter 1
Green Fluids from Vegetable Oil: Power Plant

Abstract Biodegradability has become one of the most important design parameters both in the selection of base oil and in the overall formulation of the finished fluids for insulating liquid and lubricants at electrical power generation in the power plants. There is a continuing trend toward the use of "environmentally friendly" or more readily biodegradable fluids. Readily biodegradable fluids are one that breaks down in the environment at a specified time when evaluated by standard biodegradability tests; the fluids convert to lower molecular weight components that have essentially no environmental impact. Due to the depleting of petroleum resources and environmental concern, the demand for vegetable oil-based natural esters has increased as well. In this regard, the statistical development, degradation capability of vegetable oil (triglycerides) as lubricants and liquid insulators, and the properties of natural ester-based vegetable oil in the power plant have been investigated in this chapter.

Keywords Vegetable oil • Power plants • Lubricants • Insulating fluids • Natural esters • Triglycerides • Green fluids

1.1 Introduction

Vegetable oils are triglycerides of fatty acids generally extracted from plants. Vegetable oils mostly consist of natural esters of fatty acids, i.e., triglycerides. The chemical skeleton structure of the natural esters is based on a glycerol backbone, to which three naturally occurring fatty acid groups (both saturated and unsaturated) are bonded. Again, these fatty acids may be the same or different (Fig. 1.1). Plants produce these natural esters as part of their natural growth cycle. The term "vegetable oil" can be defined as plant oil that is liquid at room temperature. Although many plant parts may yield oil, commercially, oil is extracted primarily from seeds. They are stored in the seeds and can provide a valuable high calorific foodstuff when harvested. Vegetable oilseeds have two main components, the oil part and the solid part having a protein called meal part. The production process of vegetable oil involves the removal of oil from plant parts, typically seeds. The oil

D. Gnanasekaran, V.P. Chavidi, *Vegetable Oil based Bio-lubricants and Transformer Fluids*, Materials Forming, Machining and Tribology,
DOI 10.1007/978-981-10-4870-8_1

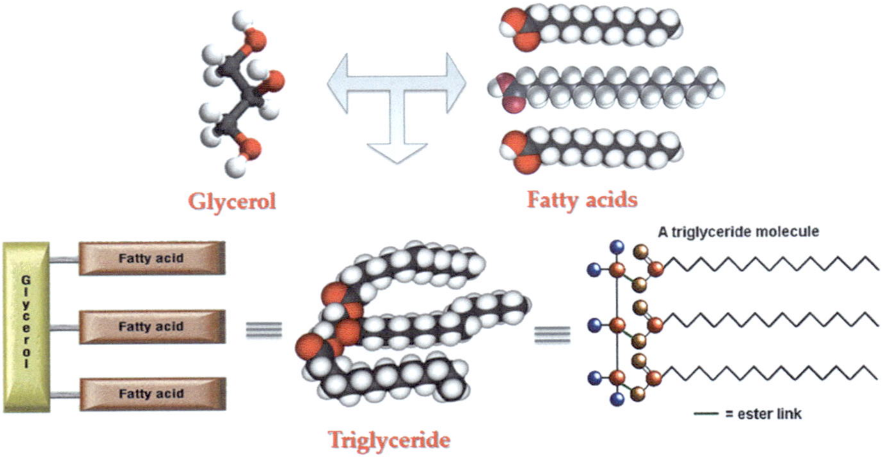

Fig. 1.1 Chemical structure of triglycerides (natural esters)

is extracted from the crude base by a process designated as RDB which stands for "refined," "bleached," and "deodorized." The commercially available RDB grade is the starting material." The complete processes are given in Sect. 6.2 and Figs. 6.5, 6.6, and 6.7.

Electrical power is one of the inevitable needs of our day-to-day life. Electrical power is majorly generated at power plants that convert mechanical energy into electrical energy. Fluids play a significant role in generation and transmission of electrical energy in power plants. Present-day power plants are facing plenty of problems, regarding the insulating oil, turbine oil, and hydraulic oils, to achieve maximum oxidation stability, superior low-temperature performance, low volatility, low toxicity, and improved additive response, which are difficult to achieve through conventional mineral oil and synthetic oils. The disposal and clearance after equipment failure and spillage are also a very difficult exercise. The leakage and spillage of mineral oil from the turbine, hydraulic, and transformer can pose a serious threat to the environment. Due to these negative points attached with mineral oil, its use is highly questionable in many countries. The use of silicone oil has some better properties like high flash point (low flammability), but they are very expensive and also nonbiodegradable. Vegetable oils are a potential substitute for petroleum-based oils; not only they are renewable, environmentally friendly, and less toxic, but also they have admirable fluid properties such as high insulating property, high lubricity, low volatility, high viscosity index, being highly available, and an alternative insulating and cooling medium for transformers. Recently, vegetable oil-based transformer fluids are increasingly replacing mineral oil-based products in the market (Chaps. 6 and 7). They are successful because they perform better than mineral oil products and they provide definite environmental and safety chains. The most important key parameter is plenty of vegetables are available in India. Figure 1.2 graphically

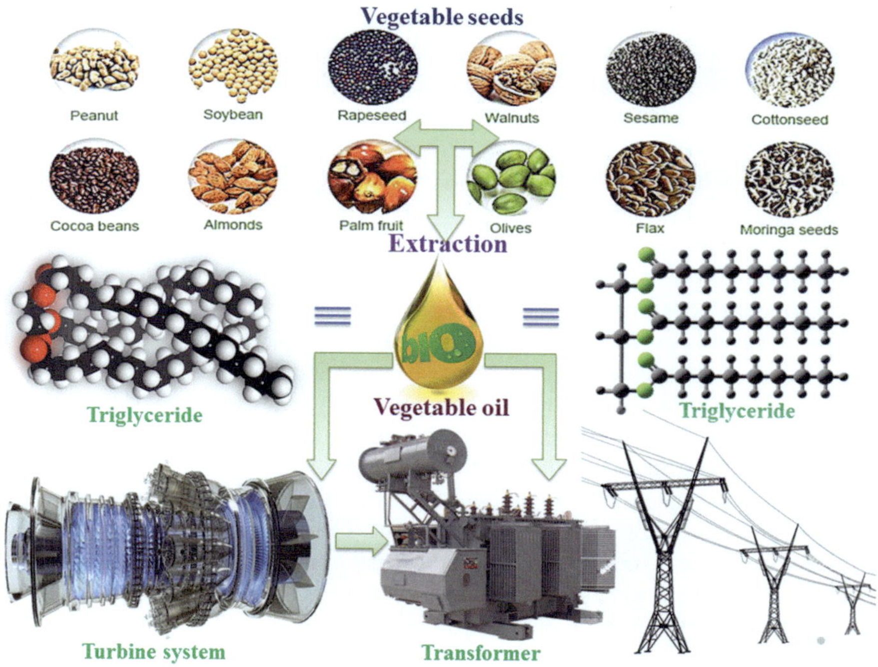

Fig. 1.2 The general process of electricity production *via* vegetable oil as fluids in power plants

represents the simple conversion process of vegetable seeds to vegetable fluid which is used in electricity generation and transmission.

Stronger environmental concerns and growing regulations over pollution through contamination will increase the need for renewable and biodegradable fluids from vegetable oils. Nowadays, there is an increasing concern about the environmental impact of oils entering the environment either by spillage during use or improper disposal of the used oil. Several countries and its organizations have made legislation to promote the usage of eco-friendly oils instead of mineral and synthetic oils, particularly in environmentally sensitive areas [1, 2]. Hence, there are trends toward better ecologically and toxicologically safe lubricants during the past decades [3]. Generally, lubricants contain some chemical additives which are used to either enhance an already existing useful property or impart desirable new properties. Most of the additives used are synthetic ester based [4, 5] and harmful to the environment and costly too. The use of vegetable oils, i.e., triglycerides of long-chain carboxylic acids combined with glycerol, as additive and base oil is highly expected from the viewpoint of decreasing global environmental pollution (see Chap. 2). The application of vegetable oil-based greener additives and base oils on the formulation of bio-lubricants has fascinated considerable attention due to their biocompatibility and enhanced multifunctional performances compared to conventional additives and mineral oil, respectively (see Chap. 2). Currently, chemically or genetically modified vegetable oils [6, 7] are used to formulate biodegradable lubricants. But their application as a base fluid is still not widespread due to economic reasons and their insufficiency to meet bulk demands. These vegetable oils can also be used as additives [8–10] as described in Chaps. 3 and 4 in the formulation of bio-lubricants, and their application as environmentally benign multifunctional additives not only increases the lifetime of engines but also increases its field service performances.

Thus, there is a greater opportunity to work in this area to develop the ecological kind of lube base oil and additives as well as transformer oil with better performance than the conventional mineral and synthetic fluids. From the literature report, it is essential to commence a methodical study on the synthesis, characterization, and performance assessment of insulating fluids (see Chaps. 6 and 7), lubricating oil (see Chaps. 2, 3, 4, and 5), and some polymeric additives (see Chaps. 3 and 4) [mainly of pour point depressant (PPD) and viscosity index improver (VII)] of vegetable oils (sunflower oil, soybean oil, etc.). There exist a lot of references to the use of modified vegetable oils as lube oil additive [11–13]. Sulfurized vegetable wax esters were designated in US Patent 4152278 as antiwear, friction modifier, and extreme pressure additives. The synthesis and characterization of vegetable oil lubricant additives, which can be used as thermal oxidative stability enhancers and viscosity improvers, were discussed in US Patent 5229023. US Patent 4873008 described the synthesis of jojoba oil-based lube oil additives. Biresaw et al. [14] have published the application of biobased polyesters as an extreme pressure additive in mineral oil. In their earlier publications, the use of vegetable oil as viscosity index improver and pour point depressant has been discussed [15–17]. Electronized vegetable oils were used as an additive in mineral base oil to enhance the extreme pressure property in the formulation of metalworking lubricants [18]. Extreme pressure (EP) additives are very important constituents for most industrial lubricant formulations that shield equipment from wear and seizure and allow it to function smoothly under heavy loads, high temperature, and low speeds [19]. Commercially available additives, i.e., alkyl and aryl disulfides and polysulfides, sulfurized hydrocarbons, fats, oils, fatty carboxylic acids, dithiocarbamates, chlorinated hydrocarbons, etc., are utilized to provide extreme protection [15]. Unfortunately, these additives are eco-toxic and not friendly with the surroundings. To overcome these limitations, Li et al. [20] have attempted to check the environmentally friendly nature of natural garlic oil as a high-performance and extreme pressure additive in lubricating applications.

1.2 Chemical Structure of Vegetable Oil and Its Suitability as Lubricants

Vegetable oils are liquid products and are extracted from plants and cash crops. Vegetable oils are primarily triglycerides, i.e., tri-esters of long-chain saturated and unsaturated fatty acids combined with glycerol, and the detailed chemical structure

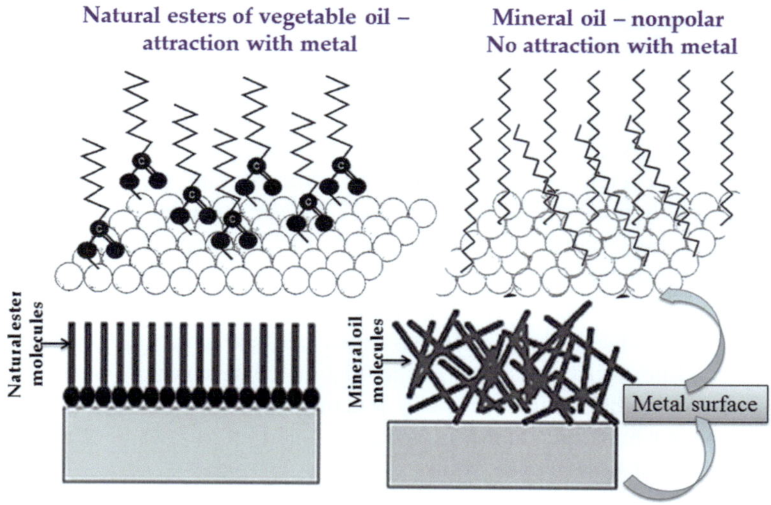

Fig. 1.3 The difference in attraction of metal with mineral and vegetable fluids [29]

of triglycerides is shown in Fig. 1.1. Such as oleic acid amide, organic straight-chain acids with polar end groups are broadly used as friction-modifying additives in the lubrication process. The fatty acids of oleic acid amide are adsorbed from hydrocarbon solutions on a metal surface. It is commonly assumed that these adsorption layers are accountable for the enhanced lubricating characteristics of oils having those substances [21]. Daniel [22] and Kipling [23] have suggested that long-chain fatty acids and alcohols from monomolecular layers of metals. The molecules are adsorbed with the major axis perpendicular to the surface, and the graphical information is given in Fig. 1.3. Recently, research has been conducted on vegetable-based metalworking fluid on the grinding performance of the various types of carboxylic acids. The fatty acid has shown the best grinding performance of all oiliness agents, friction modifiers, and EP agents. Many investigations relate to the chain length of the carbon atom of fatty acids on the lubricating properties [24–26]. There has been an increasing awareness in the use of vegetable oils for lubrication application, which has an excellent biodegradable nature and also developed the chemical stability by antiwear additive formulation [27, 28].

Mineral oils (petroleum-based oil) or synthetic hydrocarbon-type mixtures of lubricants have not met all requirements given by OEMs for lubricants utilized in various machines in different applications. Regular resolution is the blending of a few suitable additives that offer important development of base oil characteristics. General types of additives comprise polar functional groups and are appropriate to various parts of organic or organometallic compounds. The additives particularly have tribologically effective elements such as N, P, S, Cl, Zn, etc., which have capability of creating protective inorganic layers on friction surface iron or its alloys (construction material) [30]. Generally, nanomaterials, metal oxide, boron nitride, graphene, carbon nanotube, and graphite have been used as dry lubricants in high-temperature applications, where mineral lubricants are observed to be unsuitable. Presently, abundant research articles have demonstrated a sufficient control over the size, shape, and functional groups of nanoparticles [1, 31]. The potential of nanoparticles in systems produces a huge number of uses in several areas varying from tribology [32] to oil extraction [33] and from biomedical applications to drug delivery [34]. The addition of nanoparticles in lubricants may successfully decrease the interfacial friction and increase the load-bearing capacity of the parts [35]. Hence, the leading role of tribo-active antiwear (AW) or extreme pressure (EP) additives like nanomaterials is having tribologically active elements in the friction region when it is required. There are two stages of action mechanism: the first step is physical and/or chemical adsorption on a metal surface and the second step is that chemical modification takes place in the friction contact region when the wear starts. The AW and EP additives usually contain one to six tribologically dynamic elements in their monomer, and it is essential to add up to 5 wt % of additives to avoid severe metal surface exhaustion.

1.3 Scenario of Vegetable Oil as Lubricants

Triglycerides, [36] synthetic esters, and polyalkylene glycols (PAG) [37, 38] are considered as a high biodegradable fluid, which has been reported as eco-friendly fluids for tribology applications, for the reason that renewable, cheap, eco-friendly, and nontoxic, chemically or genetically revised vegetable oils are presently scrutinized as one of the most favorable resources to formulate biodegradable lubricants [11, 39–46]. The details of biodegradability of different types of base oils are shown in Table 1.1 (see Sects. 2.1 and 2.2 and Tables 2.1 and 2.2). Though, it is essential to note that the modern lubricants are formulated from a series of base fluids and additive. Therefore, the biodegradable behavior of the oil is dependent on both base fluids and additives. Conventional additives are chemically active and are more susceptible to cause water or soil pollution. Even though decomposable base oil has been broadly investigated, very few eco-friendly lubricant additives have been reported [12–14, 47]. The *Jatropha* oil has been formulated as both base oil and additive in the lubricants [19, 42]. Therefore, there is a necessity to develop

Table 1.1 Biodegradable natures of different base fluids [35, 39–47, 54, 69]

	Biodegradation nature, %	
	Methods	
Products	CEC-L-33-A-93 (21 days)	Modified Sturm (28 days)
Mineral oil	15–75%	5–50%
Synthetic esters	>55%	>40%
Vegetable oil	>90%	>70%
Readily biodegradable	87–80%	>60%

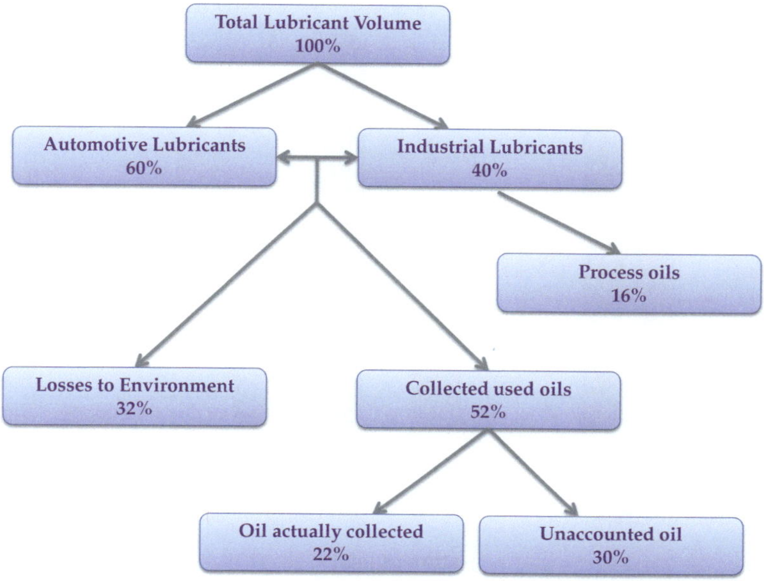

Fig. 1.4 Patterns for lubricant usage in India [48]

environmentally friendly additives as well as base fluids that are compatible with ecological rules and guidelines.

In India every year, approximately 10 million tons of mineral by-products are entering in the environments, with 40% of that representing spills, manufacturing waste and metropolitan waste, city runoff, and factory processes [48–50]. The available statistical chart is given in Fig. 1.4. Hence, strict specifications on Occupational Safety and Health Administration (OSHA), toxicity, biodegradability, and emissions have become compulsory indefinite applications. The endorsement of these specifications, along with uncertainty in the petroleum supply for political and economic reasons, has motivated the search for alternative energy sources. Vegetable oils are a viable and renewable source of environmentally favorable oils [51]. Vegetable oil is finding their way into the lubricants for industrial and transportation applications because of its properties [50, 52–58]. Waste disposal is as well of less concern for biodegradable vegetable oil-based lubricants because of their environment-friendly and nontoxic nature. As base oils, synthetic lubricants are also available and offer improved thermal and hydraulic stability and performance characteristics over refined petroleum oils, but at a superior price. Most of the biodegradable synthetic lubricants are esters that provide excellent thermal and oxidative stability [59, 60]. Figure 1.5 shows there has been developed an interest in the use of biodegradable products in the past two decades. It has been driven by environmental problems that have heightened the need to limit pollution from lubricants based on mineral oils. The plant or vegetable oils are potential alternatives to petroleum-based oils; not only they are renewable, environmentally friendly, and less toxic, but they also have excellent lubricating properties such as high lubricity,

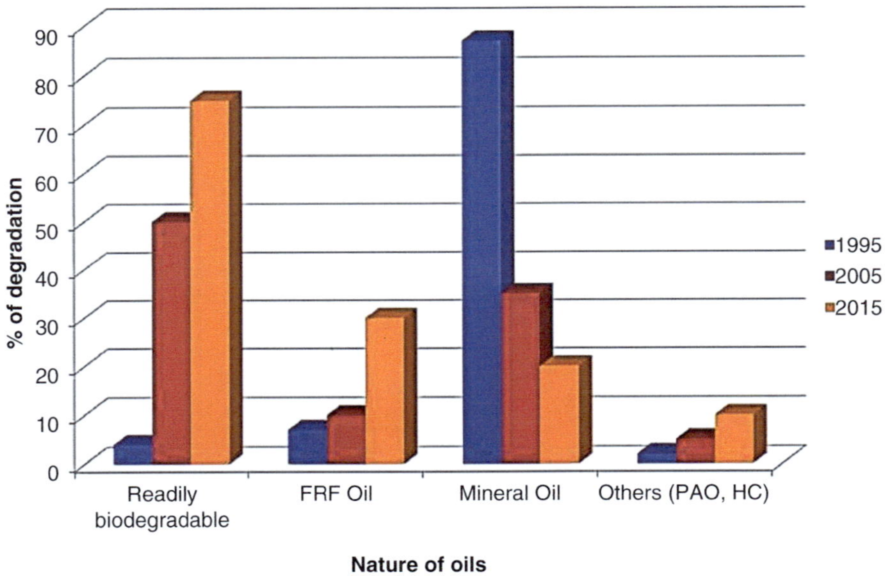

Fig. 1.5 Statistical developments of biodegradable oil in the last two decades [66–68]

low volatility, and high viscosity index [51, 52, 61–63]. The detailed lubricant's properties of vegetable oil compared with other oils are shown in Table 1.2. The chemical structure of vegetable oils, i.e., a type of fatty acid unsaturation, chain length, and branching lead to enhancement of the base oil property for lubricants. These are the main reasons the vegetable oil-based lubricants are being keenly demanded in many green industrial activities [63, 64]. The environmental and toxicity issues of mineral oil, as well as their rising cost related to a global shortage and their poor biodegradability, lead to the development of biodegradable lubricants [51, 52, 65]. The environmentally based legislation by OSHA and other international regulatory authorities discourages the use of petroleum-based mineral oil and environmentally harmful additives.

The main disadvantage of vegetable oil is easily undergoing hydrolysis, oxidation, and thermal degradation. After the genetic and chemical modification, vegetable oil can be used for various applications, even though vegetable oil has few drawbacks when used as lubricants and liquid insulators. Figure 1.6 shows the possible positions of vulnerability to thermal and chemical degradation. The cumulative details on the important application of vegetable oils and synthetic and mineral oil are shown in Table 1.3.

A laboratory test method must be meaningful, and laboratory procedures have been developed to aid in the selection of the right lubricants for the right application. It must simulate the most significant conditions of the lubricant operation. The results must be measurable and must be compared to a standard or a reference fluid. Variables that may affect the test results must be controlled. Table 1.4 provides an

Table 1.2 Oil nature *vs.* its lubrication properties [51, 52, 61–64]

Type of oils	Monounsaturated FA (%)	Polyunsaturated FA (%)	Properties of vegetable oils							
			Density (g/cm^3)	Flash point (°C)	Fire point (°C)	Pour point (°C)	Cloud point (°C)	Acid number (mg KOH/g)	Saponification number (mg KOH/lg)	Iodine number g/1g oil)
Almond	65	21	0.911	328	365	−29	−24	0.3	173	96 s
Castrol	92	5	0.957	300	358	−27	b	0.8	179	90 n
Corn	27	53	0.916	324	362	−14	−9	0.3	176	109 s
Grapeseed	16	68	0.918	324	362	−15	−11	0.7	198	150 d
Hazelnut	71	11	0.912	320	360	−13	−6	2.0	204	83 n
Linseed	16	70	0.925	320	362	−15	−14	1.4	194	192 d
Olive	76	7	0.909	318	360	−3	6	1.1	186	89 n
Peanut	45	32	0.912	326	364	−12	−3	0.3	199	105 s
Pumpkinseed	25	55	0.915	302	360	−6	b	6.1	202	100 s
Rapeseed	56	31	0.913	320	362	−29	−12	1.5	180	114 s
Safflower	12	74	0.918	328	364	−18	−12	0.4	244	154 d
Sesame	42	44	0.915	316	362	−5	−2	2.0	193	111 s
Soya been	20	67	0.917	326	364	−9	−4	0.3	195	143 d
Sunflower	13	58	0.918	325	360	−14	−6	0.3	195	130 s
Walnut	25	63	0.918	326	360	−18	−14	0.4	197	120 s
Wheat germ	16	66	0.922	316	360	−12	−6	4.9	184	130 s
Mineral	Nil	Nil	0.874	227	232	−12	−7	−b	−b	−b
Diester	Nil	Nil	0.897	273	281	−28	−23	−b	−b	−b
Tri-ester	Nil	Nil	0.916	231	239	−38	−21	−b	−b	−b

FA Fatty acid, *b* not measured, *d* drying, *s* semi-drying, *n* nondrying

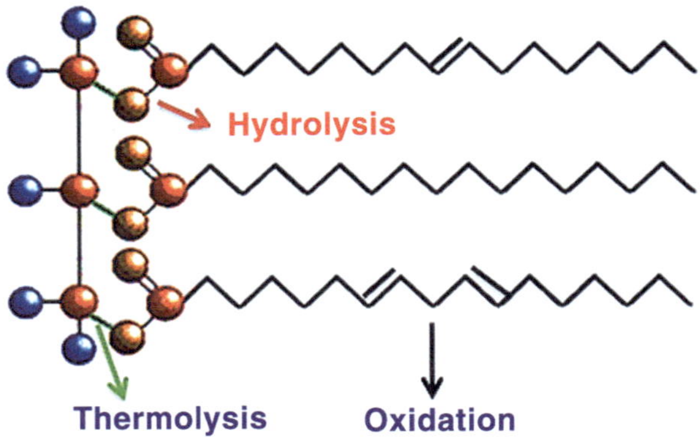

Fig. 1.6 Schematic structure of triglyceride showing positions vulnerable to degradation

overview of the methods available to conduct laboratory-scale tests for the lubricants.

The major success of vegetable oil as a base oil and additive in the lubricants is due to the presence of terminal organic functional groups which attract on the metal surface higher than the mineral oils. Figure 1.3 shows bio-based lubricants possess a polar attraction to metal, while petroleum-based fluids have no polarity and therefore no affinity to metal graphically.

Another major lubricant property of vegetable oil as a base oil and additive is its % of the fatty acid composition and its double-bond nature. Depending on the fatty acid composition of vegetable oil, the lubricating properties differ. One of the valuable properties, e.g., is an iodine value of vegetable oil which varies with the double-bond content and its fatty acid composition. The extensive discussion of vegetable oil properties against its fatty acid compositions and its double bonds is shown in Table 1.5 [81].

1.4 Scenario of Vegetable Oil as Insulating Fluids in Transformer

Oil constitutes a major component of transformers. Conventionally, mineral or synthetic oils have been used as insulating oil in the transformer for electric generation and transmission. In an attempt to overcome some of the drawbacks with mineral oils, vegetable oil has been considered as an alternative. The natural ester of vegetable oil used in the industry for several decades has proved to be a viable alternative to mineral oil and has several advantages over mineral oil (Chaps. 6 and 7). Vegetable oils are an environment-friendly liquid insulator, which does not contribute to the global greenhouse effect (they are considered a renewable energy source) and contains low levels of other pollutants. The insulating properties of the fluids are very important in power transformer, and properties of different types of fluids are compared in Tables 1.6, 7.1, 7.2, 7.3, 7.4, and 7.5. Since the beginning of the 1990s,

Table 1.3 Cumulative details of various base oils and its applications [43, 64, 69–83]

Nature of oils	Type of oils	Applications	Approx. relative costs of base oils	References
Vegetable oil	Canola oil	Hydraulic oils, tractor transmission fluids, metalworking fluids, food-grade lubes, penetrating oils, chain bar lubes	1.5–2	Shashidhara and Jayaram (2010), Bartz WJ (1998), (2006), Fox NJ, Stachowiak GW (2007), Joseph (2007), Goyan et al (1998), Lazzeri et al (2006), Lea C.W (2002), Pettersson (2007), Erhan et al (2000) and (2008), [3, 41–43, 65, 69, 71–73, 75]
	Castor oil	Gear lubricants, greases		
	Coconut oil	Gas engine oils		
	Olive oil	Automotive lubricants		
	Palm oil	Rolling lubricant, steel industry, grease		
	Rapeseed oil	Chainsaw bar lubricants, air compressor farm equipment, biodegradable greases		
	Sunflower oil – high oleic	Grease diesel fuel substitute		
	Linseed oil	Coating, paints, lacquers, varnishes, stains		
	Soybean oil	Lubricants, biodiesel fuel, metal casting/metalworking, printing inks, paints, coatings		
	Jojoba oil	Grease, cosmetic industry, lubricant applications		
Synthetic oil	*Synthetic hydrocarbons*		4–12	Asadauskas et al (1997), Masjuki et al (1990, 2003), Bhatia et al (1990), Gunderson and Hart (1962) and Randles (1999) [59, 60, 76–78]
	Alkylated aromatics	(Refrigeration) compressor oils, mineral oil-like solvency		
	Polyalphaolefins (PAO)	Engine oils, gear oils, hydraulic oils, air compressor oils		
	Polybutenes	Hydraulic oils		
	Esters			
	Diesters	Gas turbine oils, air compressor oils, hydraulic oils		
	Polyol esters	Jet engine oils, refrigeration compressor oils (chlorine-free refrigerants)		
	Phosphate esters	Fire-resistant hydraulic fluids		
	Others			
	Polyglycols			
	Silicones			
Mineral oil	Paraffins, isoparaffins	Wide range of application	1	Rudnick (2006) [64]

Table 1.4 Overview of the test methods available to conduct laboratory-scale check for lubricant properties

ASTM	Test	Procedures
D5355 & D1298	Specific gravity	To determine the specific gravity by calculating the ratio of the weight of a unit volume of the sample to the weight of a unit volume of water at 25 °C
D445	Kinematic viscosity	The kinematic viscosity is the product of the measured flow time and the calibration constant of the viscometer. The flow time can be measured by measuring the flow of a fixed volume of liquid under gravity through a capillary of calibrated viscometer at a known temperature
D2270	Viscosity index	The viscosity index is the measure of the variation in kinematic viscosity due to changes in the temperature of a lubricant product between 40 and 100 °C
D974 or D664	Total acid number	The amount of KOH required to neutralize the acid in 1 g of oil
D5558	Saponification value	Fats and oils are saponified with alcoholic KOH by boiling the solution gently and steadily. The resulting solution with indicator will be titrated against hydrochloric acid
D5554	Iodine value	A measure of the unsaturation of fats and oils and is expressed in terms of the number of centigrams of iodine absorbed per gram of sample
D5555	Free fatty acid value	The desired amount of sample is mixed with ethyl alcohol and indicator. The solution is titrated against alkali to obtain the amount of free fatty acids in the sample
D1747	Refractive index	The method to measure the refractive indexes of transparent and light-colored viscous hydrocarbon liquids and melted solids that have refractive indexes in the range between 1.33 and 1.60 and at temperatures from 80 to 100 °C
D4377	Water content	In this method, homogenized oil with solvent will be titrated to an electrometric end point using Karl Fischer reagent, the range from 0.02 to 2%
D92	Flash and fire points	Method finds out the flash and fire point in Cleveland open cup tester by heating the 70 ml of specimen rapidly at the beginning and at a slower constant rate as the flash point is approached
D97	Pour point	The lowest temperature at which movement of the specimen is observed is recorded as the pour point
D1401	Water separability	The time required for water separation is recorded for every 5 min or at some specified time limit
D892	Foam tests	Intends to determine the foaming tendency of oils by blowing air to the sample at a constant rate for 5 min and then allowing to settle for 10 min at 24 °C
D2619	Hydrolytic stability	Provides the information about the relative stability of hydraulic fluids in the presence of test conditions. Hydrolytic stability will be determined by measuring viscosity, weighting insoluble, and weight of copper strips

(continued)

Table 1.4 (continued)

ASTM	Test	Procedures
D2070	Thermal stability	The method uses a beaker containing test oil, with copper and iron rods placed in an aluminum block in an electric gravity convection oven for 168 h at a test temperature of 135 °C. The result is obtained by determining the discoloration of copper and iron rods and by measuring the quantity of sludge formed
D2272	Oxidation stability	Provides a method to determine the oxidation stability of the oils by rotating the pressure vessel
D4172	Lubricity tests	The method determines the wear-preventive characteristics of lubricating oil. The test is performed on a four-ball tester machine with the force of 392 N
D5864	Biodegradability and eco-toxicity	The method measures the biodegradation by quantifying the percentage of carbon dioxide produced by lubricant sample, which is kept under an anaerobic aquatic condition with microorganisms
D665	Rust test	Test method for rust-preventing characteristics of inhibited mineral oil in the presence of water
D3427	Air release value	Evaluates oil's capacity to release entrapped air at 25, 50, 75°C; reporting time requires releasing the entrapped air in minutes
D5185	Metal analysis	Test method for multielement determination of used and unused lubricating oils and base oils by inductively coupled plasma atomic emission spectrometry (ICP-AES)
D130	Copper strip corrosion test	Test method for corrosiveness to copper from petroleum products by copper strip test
D4927	Wear metal analysis	Standard test methods for elemental analysis of lubricant and additive components – barium, calcium, phosphorus, sulfur, and zinc by wavelength-dispersive X-ray fluorescence spectroscopy
D7873	Oxidation stability	Test method for determination of oxidation stability and insoluble formation of inhibited turbine oils at 120 °C without the inclusion of water (dry TOST method)
D2896	Total base number	Standard test method for base number of petroleum products by potentiometric perchloric acid titration

because of increasing environmental concerns, companies have started to develop vegetable oils as insulating fluids. This natural ester i.e., triglycerides, fluids have been commercially used since 1999 [84–86]. Plenty of research work has been published and some of the research works are tabulated in Table 1.7.

Natural esters are produced from vegetable oils, which are themselves manufactured from renewable (sustainable) plant crops. Natural esters offer the advantage of high fire point (Table 1.8) as well as good biodegradability, but all types of natural esters suffer from not being oxidation stable as other types of insulating liquids. The comparative class of fire and flash point with different fluids is shown in Tables 1.8 and 7.1. Though a natural ester of vegetable oils could be fashioned from a wide

Table 1.5 Vegetable oil properties *vs.* its fatty acid compositions and its double bonds [84]

Vegetable oils	Double bonds[a]	Iodine value [b]/mg per 100 g	Fatty acids (%)				
			Palmitic	Stearic	Oleic	Linoleic	Linolenic
Palm	1.7	44–58	42.8	4.2	40.5	10.1	–
Olive	2.8	75–94	13.7	2.5	71.1	10.0	0.6
Groundnut	3.4	80–106	11.4	2.4	48.3	31.9	–
Rapeseed	3.8	94–120	4.0	2.0	56.0	26.0	10.0
Sesame	3.9	103–116	9.0	6.0	41.0	43.0	1.0
Cottonseed	3.9	90–119	21.6	2.6	18.6	54.4	0.7
Corn	4.5	102–130	10.9	2.0	25.4	59.6	1.2
Soybean	4.6	117–143	11.0	4.0	23.4	53.3	7.8
Sunflower	4.7	110–143	5.2	2.7	37.2	53.8	1.0
Linseed	6.6	168–204	5.5	3.5	19.1	15.3	56.6

a average number of double bonds, *b* amount of iodine (mg) that reacts with the double bonds in 100 mg of vegetable oil

variety of crop oils, the natural esters for electrical applications are usually produced from *Jatropha*, soya, rapeseed, neem, sunflower oil, etc. The *Jatropha* oil is the most interesting among those species. It is a nonedible, mildly poisonous plant, is very resistant to aridity, and thus is not a competitor of food crops. Moreover, *Jatropha* plant needs minimal maintenance, thanks to its natural pesticide and fungicidal characteristics, its persistent life cycle (it can live up to 50 years), and its rugged physical strength. The yield of *Jatropha* is about 5 tons of seeds per hectare per year. *Jatropha* plantations have already been successfully taking place in many tropical and subtropical areas around the world (southeast of Asia, Africa, the northern states of the Indian subcontinent). Other crop oils such as coconut oil have been used, but their use is not as widespread as the main crop oils and generally occurs in countries where these crops are common. It is possible that in the future as the popularity of natural esters grows, other crop oils might appear on the market. List of available IEC and ASTM standard test methods to check the insulating properties of different types of transformer oil is shown in Table 1.9.

1.5 Conclusion

The conspectus of natural ester-based vegetable fluids (bio-lubricants and bio-insulating oils) in power plants has been enlightened in the book. In many countries, researchers and scientists have been carrying out plenty of experimental research work using vegetable oils as conventional transformer fluids and lubricant substitutes. Vegetable oil has excellent biodegradable, nontoxic, renewable nature of lubricants and insulating fluids due to its natural ester structural arrangements. In a power plant, the use of vegetable oil as transformer fluid and lubricants can play a

Table 1.6 List of insulating properties of different types of transformer oil [85, 86]

Properties	Mineral oils	Silicon oils	Synthetic esters	Vegetable oils	Test methods
Dielectric breakdown, kV	30–35	35–60	45–70	82–97	IEC 60156
Relative permittivity at 25 °C	2.1–2.5	2.6–2.9	3.0–3.5	3.1–3.3	IEC 60247
Viscosity at 0 °C, cSt	<76	81–92	26–50	143–77	ISO 3104
at 40 °C, cSt	3–16	35–40	14–29	16–37	
at 100 °C, cSt	2–2.5	15–17	4–6	4–8	
Pour point, °C	−30 to −60	−50 to −60	−40 to −50	−19 to −33	ISO 3016
Flash point, °C	100–170	300–310	250–270	315–328	ISO 2592
Fire point, °C	110–185	340–350	300–310	350–360	(1)
Density at 20 °C, kg/dm³	0.83–0.89	0.96–1.10	0.90–1.00	0.87–0.92	ISO 3675
Specific heat, $Jg^{-1}k^{-1}$	1.6–2.0	1.5	1.8–2.3	1.5–2.1	ASTM E1269
Thermal conductivity, $W\ m^{-1}\ k^{-1}$	0.11–0.16	0.15	0.15	0.16–0.17	DCS
Expansion coefficient, $10^{-4}\ K^{-1}$	7–9	10	6.5–10	5.5–5.9	ASTM D1903
Interfacial tension (IFT), dynes/cm	40–45	25	–	25	–
Moisture content, ppm dry oil	10–25	50	–	50-100	–
Heat capacity, cal/g °C	0.488	0.363	–	0.50–0.57	–
Chemical type	Hydrocarbon	Organo-silicon	–	Ester	–
Dielectric constant at 25 °C	2.2	2.71	–	3.1	–
Volume resistivity at 25 °C, ohm. cm	10^{14} –10^{15}	10^{14}	–	10^{14}	–
Breakdown voltage, kV, ASTM D1816, 2 mm gap electrodes	60	–	–	74	–
Impulse breakdown voltage, kV	145	136	–	116	–
Dissipation factor (%) 25 °C 100 °C	0.05 max 0.3 max	-0.01	-	0.25–1.00	–
Grassing tendency – ASTM D2300	−10–20	N/A	–	−50	–
Biodegradability, % CEC-L-33 (21 days)	30	Very low	–	97–99	–

Table 1.7 Cumulative details of recent research work carried out on vegetable oil as liquid insulator in transformer

Properties		Achmad Susilo et al. [87]		Gomez et al. [88]				Yanttar Z. Arief et al. [89]		
		Mineral oil	PFAE	Bivolt A	Bivolt HW	FR3	Mineral	Palm oil	Coconut oil	Sunflower oil
Electrical	Dielectric strength (kV)	70–75/2.5 mm	85	50	50	48	57	75	60	38–45
	Relative permittivity	2.2@80 °C	2.95@80 °C	–	–	–	–	3.1@40 °C	2.79@20 °C	3.1@25 °C
	Tan δ	0.001@40 °C	0.31@80 °C	1.6 @100 °C	0.77 @100 °C	1.9 @100 °C	0.08 @100 °C	0.03@25 °C	0.08@20 °C	0.0093@25 °C
	Volume resistivity	7.6×10^{15}	7.1×10^{12}	–	–	–	–	–	–	–
Physical	Viscosity, 40 °C									
	mm^2/s	8.13	5	–	–	–	–	300@25 °C	29.8–31.6	43
	cSt	–	–	36.6	40.1	36.7	10	–	–	–
	Flash point, °C	152	176	308	308	314	138	>220	225	<330
	Fire point, °C	–	–	342	338	338	148	–	–	–
	Pour point, °C	–45	–32.5	–	–	–	–	–	–	–
	Density									
	g/cm^3	0.88@40 °C	0.86@40 °C	–	–	–	–	–	–	–
	kg/m^2	–	–	–	–	–	–	0.9@15 °C	0.92@20 °C	0.919@20 °C
Environ-mental	TAN (mgKOH/g)	0.04	≤0.03	–	–	–	–	0.07	0.02	0.02
	Toxicity	<0.01	0.05	–	–	–	–	–	–	–
	Biodegradability	–	–	–	–	–	–	–	–	–
	Source	Crude oil	Palm oil	Corn flower	Sunflower	Soya oil	Petroleum	Palm oil	coconut	sunflower
	Appearance	–	–	–	–	–	–	–	–	–

Properties		Kiyoshi takamoto et al.			Villarroel et al. [90]		Shuhan yaol et al. [91]		Zhengjiand wang et al. [92]	
		PFAE Oil	Rapeseed oil	Mineral oil	Bio electra	Bio temp	geminix	FP3	Cammellia oil	Mineral oil
Electrical	Dielectric strength (kV)	81/2.5 mm	74/2.5 mm	70–75/2.5 mm	65	65	≥35	56	–	–
	Relative permittivity	2.95@80 °C	2.86@80 °C	2.2@80 °C	–	–	2.2@90 °C	3@90 °C	–	–
	Tan δ	3.1×10^{-3} @80 °C	8.3×10^{-2} @80 °C	1×10^{-3} @80 °C	–	–	≤0.1@90 °C	4×10^{10} @90 °C	–	–
	Volume resistivity	7.1×10^{12}	4.4×10^{12}	7.6×10^{15}	–	–	1×10^{12} @90 °C	4×10^{10} @90 °C	–	–
Physical	Viscosity, @40 °C									
	mm^2/s	5.06	36	8.13	–	–	≤13	≤34.1	39.9	≤13
	cSt	–	–	–	39.2	45	–	–	–	–
	Flash point, °C	176	334	152	330	330	≥135	316	322	>135
	Fire point, °C	–	–	–	–	–	–	–	–	–
	Pour point, °C	–32.5	–27.5	–45	–26	–15	<–22	–21	–28	<–22
	Density,@20 °C	–	–	–	–	–	–	–	–	–
	g/cm^3	0.86	0.92	0.88	–	–	–	–	–	–
	kg/m^2	–	–	–	–	–	<0.895	0.91	0.90	≤1.3
Environ-mental	TAN (mgKOH/g)	–	–	–	–	–	≤0.03	0.04	0.04	≤0.03
	Toxicity	Nontoxic	Nontoxic	Slight toxic	–	–	–	–	–	–
	Biodegradability	high	high	low	–	–	–	–	–	–
	source	Palm oil	Rapeseed	Crude oil	Sunflower seed	Sunflower seed	–	Soyabeen oil	–	Crude oil
	Appearance	–	–	–	–	–	Transparent	Light green	Light yellow	Transparent

PFAE: palm fatty acid ester, Biovolt A: from corn oil, Biovolt HW: from sunflower oil with high oleic acid, Envirotemp FR3: from soya oil

Table 1.8 Insulating properties of different types of transformer oils

Fluid type	Flash point, °C	Fire point, °C	Class	Class	Net calorific value	Ester linkages	Approx. water absorption at 23 °C
Mineral oil	160–170	170–180	O	1	≥42MJ/Kg	0	55
Silicon oil	>300	>350	K3	3	≥32MJ/Kg	0	220
Natural ester	>300	>350	K2	2	≥42MJ/Kg and <32MJ/Kg	3	1100
Synthetic ester	>250	>300	K3	3	<32MJ/Kg	4	2600

Note: O Class ≤ 300 °C, K Class >300 °C according to IEC 61100

Table 1.9 List of available IEC, ASTM standard: insulating properties vs. different fluid testing methods

Properties	Mineral oil [93]	Synthetic ester [94]	Natural ester [95]	Silicone fluid [96]
Acidity	**IEC 62021-1/IEC 62021-2**, *ASTM D974*	**IEC 62021-1/ IEC 62021-2**	*ASTM D974*	**IEC 62021-1**
Appearance	ISO 2049	**ISO 2049**	*ASTM D1524*	**Visual, ISO 2049**
Breakdown voltage	**IEC 60156**, *ASTM D1816*	**IEC 60156**	*ASTM D877, ASTM D1816*	**IEC 60156**
Color	ISO2049, **ISO2211,** *ASTM D1500*	**ISO 2211**	*ASTM D1500*	**ISO 2211**
Corrosive sulfur	IEC 62535, *ASTM D1275*	–	*ASTM D1275*	–
Dielectric dissipation factor (DDF)	**IEC 60247/IEC 61620** *ASTM D924*	**IEC 60247**	*ASTM D924*	**IEC 60247**
Density	**ISO 3675/ISO 12185** *ASTM D1298*	**ISO 3675**	*ASTM D1298*	**ISO 3675**
DGA analysis	IEC 60567	–	**ASTM D2945, ASTM D3284, ASTM D3612**	–
Fire point	ISO 2592, *ASTM D92*	**ISO 2592**	*ASTM D92*	**ISO 2592**
Flash point	**ISO 2719**, ISO 2592, *ASTM D92*	**ISO 2719/ ISO 2592**	*ASTM D92*	**ISO 2719/ISO 2592**
Furanic compounds	**IEC 61198**, *ASTM D5837*	–	–	–

(continued)

Table 1.9 (continued)

Properties	Mineral oil [93]	Synthetic ester [94]	Natural ester [95]	Silicone fluid [96]
Gassing tendency	**IEC 60628**, *ASTM D2300*	**IEC 60628**	*ASTM D2300*	**IEC 60628**
Interfacial tension	**ISO 6295**, *ASTM D971*	–	*ASTM 971*	*ASTM 971*
Kinematic viscosity	**ISO 3104, ASTM D455**	**ISO 3104**	**ASTM D455**	**ISO 3104**
Kinematic viscosity at low T	**IEC 61868**	–	–	–
Lightning impulse breakdown	IEC 60897, *ASTM D3300*	–	*ASTM D3300*	–
Oxidation stability	**IEC 61125**, IEC 62036 *ASTM D2112, ASTM D2440*	**IEC 61125**	–	–
PCB content	**IEC 61619**, *ASTM D4059*	–	**ASTM D4059**	–
Permittivity	**IEC 60247**, *ASTM D924*	**IEC 60247**	*ASTM D924*	**IEC 60247**
Pour point	**ISO 3016**, *ASTM D97*	**ISO 3016**	*ASTM D97*	**ISO 3016**
Refractive index	ISO 5661	**ISO 5661**	–	ISO 5661
Resistivity	**IEC 60247**	**IEC 60247**	ASTM D1169	**IEC 60247**
Specific heat	*ASTM D2766*	–	*ASTM D2766*	–
Stray gassing	CIGRE Brochure n°296	–	–	–
Thermal conductivity	*ASTM D2717*	–	*ASTM D2717*	–
Thermal expansion coef.	*ASTM D1903*	–	*ASTM D1903*	–
Visual examination	*ASTM D1524*	–	*ASTM D1524*	–
Water content	**IEC 60814**, *ASTM D1533*	**IEC 60814**	*ASTM D1533*	**IEC 60814**

Key: **most commonly used IEC method**, *most commonly used ASTM method*, not quoted but generally used

vital role in helping the world to reduce the environmental impact of mineral and synthetic oils. The development of vegetable fluid satisfies the current requirements for environmentally friendly fluid for insulating oil and lubricants. However, the environmental benefits are of essential current interest; in the future, petroleum products are eventually going to exhaust by the mid-twenty-first century, and there would be serious lacks of these products. Luckily, the groundwork has been arranged already by the development of suitable insulating liquid and lubricants. Vegetable oils are naturally suitable to be used as fluids in lubricating and insulating; with the cooperation of chemical modification to enhance its properties, the vegetable fluids are functioning as well as the mineral and synthetic oils or better. Actually, green fluids (biodegradable lubricants and transformer oil, refined from vegetables) are those that optimize energy efficiency and minimize wear in the machinery which they lubricate and have maximized service lifetimes with reduced quantity of fluids required. Hence, for the job, there is a lot of scope for the development of vegetable-based fluid technology in the power sector, but lubricants and insulating fluids have a significant contribution to make energy conservation, minimization of waste, and development of durable products. These principles are expected to lead to increased use of high-performance vegetable-based fluids as effective additives. Green fluids will have an important role to play in the design and formulation of novel materials required for the future.

References

1. Rizvi SQ (2009) Lubricant additives. In: A comprehensive review of lubricant chemistry, technology, selection, and design. ASTM International, West Conshohocken
2. Betton CI (2010) Lubricants & their environmental impact. In: Mortier RM, Fox MF, Orszulik ST (eds) Chemistry and technology of lubricants. Springer, Dordrecht, pp 435–457
3. Pettersson A (2007) High-performance base fluids for environmentally adapted lubricants. Tribol Int 40(4):638–645
4. Abdel-Azim AAA, Nassar AM, Ahmed NS, Kamal RS (2006) Preparation and evaluation of acrylate polymers as pour point depressants for lube oil. Pet Sci Technol 24(8):887–894
5. Ghosh P, Das M, Upadhyay M, Das T, Mandal A (2011) Synthesis and evaluation of acrylate polymers in lubricating oil. J Chem Eng Data 56:3752–3758
6. Wu X, Zhang X, Yang S, Chen H, Wang D (2000) The study of epoxidized rapeseed oil used as a potential biodegradable lubricant. J Am Oil Chem Soc 77:561–563
7. (a). Sharma BK, Adhvaryu A, Erhan SZ (2006) Synthesis of hydroxyl thio-ether derivatives of vegetable oil. J Agric Food Chem 54(26):9866–9872. (b). Mohammed Aji, Shettima Abba Kyari, Gideon Zoaka (2015) Comparative Studies between bio lubricants from Jatropha, neem oil and mineral lubricant (Engen Super 20w/50). Appli Rese J 1(4):252–257
8. Maleque MA, Masjuki HH, Sapuan SM (2003) Vegetable-based biodegradable lubricating oil additives. Ind Lubr Tribol 55(3):137–143
9. Ertugrul D, Filiz K (2004) Using cotton seed oil as environmentally accepted lubricant additive. Energy Sources 26:611–625
10. Kammann KP Jr, Phillips AI (1985) Sulfurized vegetable oil products as lubricant additives. J Am Oil Chem Soc 62:917–923
11. Fuks IG, Evdokimov AY, Dzhamalov AA, Luksa A (1992) Vegetable oils and animal fats as raw materials for the manufacture of commercial lubricants. Chem Technol Fuels Oils 28(4):230–237
12. Sharma BK, Adhvaryu A, Erhan SZ (2006) Synthesis of hydroxyl thio-ether derivatives of vegetable oil. J Agric Food Chem 54(26):9866–9872
13. Ghosh P, Karmakar G (2013) Green additives for lubricating oils. ACS Sustain Chem Eng 1(11):1364–1370
14. Biresaw G, Asadauskas SJ, McClure TG (2012) Polysulfide and biobased extreme pressure additive performance in vegetable vs paraffinic oils. Ind Eng Chem Res 51:262–273
15. Papay AG (1998) Antiwear and extreme pressure additives in lubricants. Lubr Sci 10(3):209–224

16. Chin-Chung W, Lee-Yan S, Haw-Wen C (2001) Differential effects of garlic oil and its three major organosulfur components on the hepatic detoxification system in rats. J Agric Food Chem 50(2):378–383
17. Weinberg DS, Manier ML, Richardson MD, Haibach FG (1993) Identification and quantification of organosulfur compliance markers in a garlic extract. J Agric Food Chem 41(1):37–41
18. Braun E, Buyanovskii I, Pravotorova E (2007) Reduction of amount of tribotests of lubricants with four-ball machine. J Frict Wear 28(3):300–305
19. a. Thomas R, Achim F (2009) Sulfur carriers. In: Rudnick LR (ed) Lubricant additives: chemistry and applications. CRC Press, Boca Raton. b. Imran A, Masjuki HH, Kalam MA, Varman M, Hasmelidin M, Al Mahmud KAH, Shahir SA, Habibullah M (2013) Study of friction and wear characteristic of Jatropha oil blended lube oil. Procedia Eng 68:178–185
20. Li W, Jiang C, Chao M, Wang X (2014) Natural garlic oil as a high performance, environmentally friendly, extreme pressure additive in lubricating oils. ACS Sustain Chem Eng 2:798–803
21. Studt P (1989) Boundary lubrication: adsorption of oil additives on steel and ceramic surfaces and its influence on friction and wear. Tribol Int 22:111–119
22. Daniel SG (1951) The adsorption on metal surfaces of long-chain polar compounds from hydrocarbon solutions. Trans Faraday Soc 47:1345–1359
23. (a). Kipling JJ (1965) Adsorption from solutions of non-electrolytes. Academic Press, London, p 55. (b). Bowman WF, Stachowiak GW (1999) Application of sealed capsule differential scanning calorimetry - part II: assessing the performance of antioxidants and base oils. Lubr Eng 55(5):22–9
24. Hironaka S, Yahagi Y, Sakurai T (1975) Heats of adsorption and anti-wear properties of some surface active substance. Bull Jpn Petrol Inst 17(2):201–205
25. Jahanmir S, Beltzer M (1986) An adsorption model for friction in boundary lubrication. ASLE Trans 29(3):423–430
26. Nakayama K, Studt P (1991) Additive interaction and lubrication performance in polar additive binary system. Tribol Int 25:2
27. Arnsek A, Vizintin J (1999) Scuffing and load capacity of rapeseed-based oils. Lubr Eng 55:11–18
28. Gunsel S, Lockwood FE (1995) The influence of copper-containing additives on oil oxidation and corrosion. Tribol Trans 38:485–496
29. Bob F, Dwight S (2010) The benefits of bio-based lubricants. Available in pdf from Gear Solutions: http://www.gearsolutions.com/article/detail/5990/the-benefits-of-bio-based-lubricants
30. Bakunin VN, Yu Suslov A, Kuzmina GN, Parenago OP, Topchiev AV (2004) Synthesis and application of inorganic nanoparticles as lubricant components – a review, Institute of Petrochemical Synthesis RAS, Moscow, Russia. J Nanopart Res 6:273–284
31. Choi SUS (1995) Enhancing thermal conductivity of fluids with nanoparticles. In: Siginer DA, Wang HP (eds) Developments and applications of non-Newtonian flows. ASME, New York, pp 99–105
32. (a). Ji X, Chen Y, Zhao G, Wang X, Liu W (2011) Tribological properties of CaCO nanoparticles as an additive in lithium grease. Tribol. Lett 41:113–119. (b). Friedman H, Eidelman O, Feldman Y, Moshkovich A, Perfiliev V, Rapoport L, Cohen H, Yoffe A, Tenne R (2007) Fabrication of self-lubricating cobalt coatings on metal surfaces. Nanotechnol 18 doi:10.1088/0957-4484/18/11/115703
33. Chaudhury MK (2003) Complex fluids: spread word about nanofluids. Nature 423:131–132
34. Kleinstreuer C, Li J, Koo J (2008) Microfluidics of nano-drug delivery. Int J Heat Mass Transf 51:5590–5597
35. (a). Rapoport L, Fleischer N, Tenne R (2003) Fullerene-like WS nanoparticles: superior lubricants for harsh conditions. Adv Mater 15:651–655. (b). Cumings J, Zettl A (2000) Low-friction nanoscale linear bearing realized from multiwall carbon nanotubes. Sci 289:602–604. (c). Falvo MR, Taylor RM, Helser A, Chi V, Brooks FP Jr, Washburn S, Superfine R (1999) Nanometre-scale rolling and sliding of carbon nanotubes. Nature 397:236–238. (d) Akbulut

M, Belman N, Golan Y, Israelachvili J (2006) Frictional properties of confined nanorods. Adv Mater 18:2589–2592
36. Salimon J, Salih N, Yousif E (2010) Biolubricants: raw materials, chemical modifications, and environmental benefits. Eur J Lipid Sci Technol 112(5):519–530
37. Hahn S, Dott W, Eisentraeger A (2006) Characterization of aging behavior of environmentally acceptable lubricants based on trimethylolpropane esters. J Synth Lubr 23(4):223–236
38. Pettersson A (2003) Tribological characterization of environmentally adapted ester based fluids. Tribol Int 36(11):815–820
39. Hashem AI, Abou Elmagd WSI, Salem AE, El-Kasaby M, El-Nahas AM (2013) Conversion of some vegetable oils into synthetic lubricants. Energy Sources, Part A 35(5):397–400
40. (a). Fox NJ, Stachowiak GW (2007) Vegetable oil-based lubricants, a review of oxidation. Tribol Int 40(7):1035–1046. (b). Steve B (2002) Green lubricants. Environ Benefits Impacts Lubr Green Chem 4:293–307
41. Erhan SZ, Sharma BK, Liu Z, Adhvaryu A (2008) Lubricant base stock potential of chemically modified vegetable oils. J Agric Food Chem 56(19):8919–8925
42. (a). Erhan SZ, Asadauskas S (2000) Lubricant base stocks from vegetable oils. Ind Crops Prod 11(2–3):277–282. (b). Syaima MTS, Zamratul MIM, Noor IM, Rifdi WMWT (2014) Development of bio-lubricant from Jatropha curcas oils. IJRCMCE 1:10–12
43. Shashidhara YM, Jayaram SR (2010) Vegetable oils as a potential cutting fluid-an evolution. Tribol Int 43(5–6):1073–1081
44. Sharma BK, Liu Z, Adhvaryu A, Erhan SZ (2008) One-pot synthesis of chemically modified vegetable oils. J Agric Food Chem 56(9):3049–3056
45. Campanella A, Rustoy E, Baldessari A, Baltanás MA (2010) Lubricants from chemically modified vegetable oils. Bioresour Technol 101(1):245–254
46. Ozcelik B, Kuram E, Cetin MH, Demirbas E (2011) Experimental investigations of vegetable based cutting fluids with extreme pressure during turning of AISI 304L. Tribol Int 44(12):1864–1871
47. Li W, Wu Y, Wang X, Liu W (2012) Tribological study of boron containing soybean lecithin as an environmentally friendly lubricant additive in synthetic base fluids. Tribol Lett 47(3):381–388
48. (a). Mohanti S (2001) Envirotech New Delhi www.envirotechindia.in/ (b). Amit Suhane, Sarviya RM, Rehman A, Khaira HK (2014) Prospects of vegetable oils for lubricant formulations in Indian perspectives. IJETR 2(12):2321–0869
49. Gawrilow I (2004) Vegetable oil usage in lubricants. Oleochem Inform 15:702–705
50. Sharma BK, Adhvaryua A, Liu Z, Erhan SZ (2006) Chemical modification of vegetable oils for lubricant applications. J Am Oil Chem Soc 83:129–136
51. Soni S, Agarwal M (2014) Lubricants from renewable energy sources – a review. Green Chem Lett Rev 7:359–382
52. Nagendramma P, Kaul S (2012) Development of ecofriendly/biodegradable lubricants: an overview. Renew Sustain Energy Rev 16:764–774
53. Padmaja KV, Rao BVSK, Reddy RK, Bhaskar PS, Singh AK, Prasad RBN (2012) Undecenoic acid-based polyol esters as potential lubricant base stocks. Ind Crop Prod 35:237–240
54. Arumugam S, Sriram G, Subhadra L (2012) Synthesis, chemical modification and tribological evaluation of plant oil as bio-degradable low temperature lubricant. Procedia Eng 38:1508–1517
55. Quinchia LA, Reddyhoff T, Delgado MA, Gallegos C, Spikes HA (2014) Tribological studies of potential vegetable oil-based lubricants containing environmentally friendly viscosity modifiers. Tribol Int 69:110–117
56. Singh AK (2011) Castor oil-based lubricant reduces smoke emission in two-stroke engines. Ind Crop Prod 33:287–295
57. Quinchia LA, Delgado MA, Valencia C, Franco JM, Gallegos C (2010) Viscosity modification of different vegetable oils with EVA copolymer for lubricant applications. Ind Crop Prod 32:245–254

58. (a). Campanella A, Rustoy E, Baldessari A, Baltanas MA (2010) Lubricants from chemically modified vegetable oils. Ind Crops Prod 101:245–254. (b). Bilal S, Mohammed-Dabo IA, Nuhu M, Kasim SA, Almustapha IH, Yamusa YA (2013) Production of biolubricant from Jatropha curcas seed oil. J Chem Eng Mater Sci 4:72–79
59. Gunderson RC, Hart AW (eds) (1962) Synthetic lubricants. Reinhold, New York
60. Randles SJ (1999) Esters. In: Rudnick LR, Shubkin RL (eds) Synthetic lubricants and high-performance functional fluids. Marcel Dekker, New York, pp 63–102
61. Lathi P, Mattiasson M (2007) Green approach for the preparation of biodegradable lubricant base stock from epoxidized vegetable oil. Appl Catal B 9:207–212
62. Jayadas N, Nair K (2006) Coconut oil as a base oil for industrial lubricants-evaluation and modification of thermal, oxidative and low-temperature properties. Tribol Int 39:873–878
63. Lawal SA (2013) A review of application of vegetable oil-based cutting fluids in machining non-ferrous metals. Indian J Sci Technol 6:3951–3956
64. Rudnick LR (2006) Synthetics, mineral oils, and bio-based lubricants: chemistry and technology. CRC/Taylor & Francis Group, New York
65. Lazzeri L, Mazzoncini M, Rossi A, Balducci E, Bartolini G, Giovannelli L, Pedriali R, Petroselli R, Patalano G, Agnoletti G, Borgioli A, Croce B, Avino LD (2006) Biolubricants for the textile and tannery industries as an alternative to conventional mineral oils: an application experience in the Tuscany province. Ind Crop Prod 24:280–291
66. Dietrich H (2002) Recent trends in environmentally friendly lubricants. J Synth Lubr 18:328–347
67. Liew Yun Hsien W (2015) Towards green lubrication in machining, springer briefs in green chemistry for sustainability. Springer, Malaysia
68. Samarth NB, Mahanwar PA (2015) Modified vegetable oil-based additives as a future polymeric material-review. Open J Organic Polym Mater 5:1–22
69. Bartz WJ (1998) Lubricants and the environment. Tribol Int 31:35–47
70. Bartz WJ (2006) Ecotribology: environmentally acceptable tribological practices. Tribol Int 39:728–733
71. Fox NJ, Stachowiak GW (2007) Vegetable oil-based lubricants - a review of oxidation. Tribol Int 40:1035–1046
72. Joseph PV, Deepak S, Sharma DK (2007) Study of some non-edible vegetable oils of Indian origin for lubricant application. J Synth Lubr 24:181–197
73. Goyan RL, Melley RE, Wissner PA, Ong WC (1998) Biodegradable lubricants. Lubr Eng 54(7):10–17
74. Chauhan PS, Chhibber VK (2013) Non-edible oil as a source of bio-lubricant for industrial applications: a review. Inter J Eng Sci Innov Technol 2(1):299–306
75. Lea CW (2002) European development of lubricants from renewable sources. Ind Lubr Tribol 54(6):268–274
76. Asadauskas S, Perez JH, Duda JL (1997) Lubrication properties of castor oil–potential base stock for biodegradable. Lubr Eng 53(12):35–40
77. Masjuki HH, Maleque MA, Kubo A, Nonaka T (1990) Palm oil and mineral oil based lubricants-their tribological and emission performance. Tribol Int 32:305–314
78. Maleque MA, Masjuki HH, Sapuan SM (2003) Vegetable based biodegradable lubricating oil additives. Ind Lubr technol 55(3):137–143
79. Erhan SZ, Asadauskas S (2000) Lubricant base stocks from vegetable oils. Ind Crop Prod 11:277–282
80. Bhatia VK, Chaudhry A, Sivasankaran GK, Bisht RPS, Kashyap M (1990) Modification of jojoba oil for lubricant formulations. JAOCS 67(1):1–7
81. Belgacem MN, Gandini A (2008) In: Belgacem MN, Gandini A (eds) Monomers, polymers, and composites from renewable resources. Elsevier, Amsterdam, pp 39–66
82. Bertrand Y, Hoang LC (2003) Vegetal oils as a substitute for mineral oils. In: Proceedings of the 7th international conference on properties applications of dielectric materials, Nagoya

83. Oommen TV, Claiborne CC, Walsh EJ, Baker JP (2000) A new vegetable oil based transformer fluid: development and verification, Electrical Insulation and Dielectric Phenomena, Annual Report Conference on, vol. 1. Victoria, pp 308–312
84. Oommen TV, Claiborne CC, Walsh EJ, Baker JP (1999) Biodegradable transformer fluid from high oleic vegetable oils, International Conference of Doble Clients, 1999
85. McShane P, Corkran J, Rapp K, Luksich J (2006) Natural ester dielectric fluid development, IEEE
86. Bertrand Y, Hoang LC (2004) Vegetable oils as a substitute for mineral insulating oils in medium voltage equipment, CIGRE, paper D1–202
87. Susilo A, Muslim J, Arief YZ, Muhamad NA, Hikita M, Kozako M, Tsuchie M, Suzuki T, Hatada S, Kanetani A, Kano T, Suwarno, Khayam U (2014) Comparative study of partial discharge characteristics and dissolved gas analysis on palm-based oil as insulating material. Power Engineering and Renewable Energy (ICPERE), International Conference on, Bali, pp 232–236
88. Gómez NA, Wilhelm HM, Santos CC, Stocco GB (2014) Dissolved gas analysis (DGA) of natural ester insulating fluids with different chemical compositions. IEEE Trans Dielectr Electr Insul 21(3):1071–1078
89. Arief YZ, Ahmad MH, Lau KY, Muhamad NA, Bashir N, Mohd NK, Huey LW, Kiat YS, Azli SA (2014) A comparative study on the effect of electrical aging on electrical properties of palm fatty acid ester (PFAE) and FR3 as dielectric materials. In Conference Proceeding - IEEE International Conference on Power and Energy, PECon 2014. [7062427] Institute of Electrical and Electronics Engineers Inc, pp 128–133
90. Villarroel R, García D, García B, Burgos J (2015) Moisture diffusion coefficients of transformer pressboard insulation impregnated with natural esters. IEEE Trans Dielectr Electr Insul 22(1):581–589
91. Yao S, Li J, Li L, Liao R, Zhou J (2014) Comparison analysis to thermal aging properties of vegetable and mineral insulating oils, High Voltage Engineering and Application (ICHVE), International Conference on, Poznan, 2014, pp 1–4
92. Wang Z, Li J, Yang L, Liao R (2014) Characteristics of acid value in vegetable insulating oil during thermal aging, High Voltage Engineering and Application (ICHVE), International Conference on, Poznan, 2014, pp 1–4
93. IEC 60422 Mineral insulating oils in electrical equipment supervision and maintenance guidance
94. IEC 61203 synthetic organic esters for electrical purposes- Guide for maintenance of transformer esters in equipment
95. D6871–03 (2008) Standard Specification for Natural (Vegetable Oil) Ester fluids Used in electrical apparatus
96. IEC 60944 (Ed. 1) -1988 Guide for Acceptance of Silicone Transformer Fluid and Its Maintenance in Transformers

Part II
Lubricants from Vegetable Oil

Chapter 2
Biodegradable, Renewable, and Eco-friendly Vegetable Oil: Lubricants

Abstract The environmental pollution, toxicity, and rising cost of conventional lubricants lead to renewed awareness in the improvement of environmentally friendly bio-lubricants. Due to the negative impact (low biodegradability and more toxicity of mineral and synthetic) on the environment, there has been a stable increase in the demand for biodegradable and eco-friendly bio-lubricants. Perhaps, mineral oils pollute the atmospheric air, soil, and drinking water and disturb people's life and plants to a greater level. However, the major problem leads to the exhaustion of the world's crude oil, increasing crude oil prices, and problems connected to preservation have brought about renewing or reusing awareness in the use of biodegradable lubricants. Definitely, oils of natural ester are capable as a base stock for environmentally friendly bio-lubricants due to its lubricity, biodegradation capability, viscosity vs. temperature characteristics, low evaporation capacity, etc. The chapter covers the biodegradable mechanism, toxicity ration, and eco-friendly natures of natural esters of vegetable oils.

Keywords Vegetable oil • Biodegradation • Bio-lubricants • Eco-friendly • Nontoxic • Base stocks

2.1 Introduction

2.1.1 Biodegradable Nature of Vegetable Oil

Biodegradability is defined as the capability of the material to be disintegrated by microorganisms and the method of chemical collapse or conversion of a material produced by microorganisms or enzymes. Numerous selective characterization methods are employed to evaluate the biodegradable nature of bio-lubricants and their additives. Table 2.1 gives the several countries adopted biodegradable test methods, and the Table 2.2 delivers the typical biodegradability nature of the different oils as measured by the CEC-L-33-A-94 and OECD 301B procedures. Different countries have adopted different criteria and restrictions for toxicity, and biodegradability depends on their respective sociopolitical environmental conditions. An

D. Gnanasekaran, V.P. Chavidi, *Vegetable Oil based Bio-lubricants and Transformer Fluids*, Materials Forming, Machining and Tribology,
DOI 10.1007/978-981-10-4870-8_2

Table 2.1 Biodegradable test methods and its criterion [1, 2]

	Time (hours)	Parameter measured	Criterion
Ready biodegradability			
1. Modified AFNOR OECD 301A	28	Loss of dissolved organic carbon	>70%
2. Modified Sturm OECD 301B	28	Production of CO2	>60%
3. Modified MITI OECD 301C	28	Oxygen demand	>60%
4. Closed Bottle OECD 301D	28	Oxygen demand	>70%
Inherent biodegradability			
Modified semicontinuous activated sludge (SCAS) OECD 302A	>28	Loss of dissolved organic carbon	>20%
Zahn-Wellens EMPA tests OECD 302B	28	Loss of dissolved organic carbon	>20%
Relatively primary biodegradation CEC-L-33-A-94	21	Loss of hydrocarbon infrared bands at 2930 cm^{-1}	>70% to ≥90

Table 2.2 Biodegradability nature of the different base fluids [1, 2]

Type of fluids	Biodegradability (%), CEC-L-33-A-94 method, 21 days	Ready biodegradability, OECD 301B, 28 days
Mineral oil	20–40	10–40
Vegetable oil	90–98	75–95
Esters	75–100	50–95
Polyols	70–100	0–85
Trimelliates	0–70	–

assessment of these methodologies suggests that the choice of rapidly biodegradable, nontoxic base fluids boils down to vegetable oils, vegetable oil-derived esters, and synthetic esters.

Currently, available biodegradable lubricants are five types of base oil of some significance. These are:

- Highly double bonded or high oleic vegetable oils (HOVOs)
- Low viscous polyalphaolefins
- Polyalkylene glycols
- Dibasic acid esters
- Polyol esters

This chapter focuses only on biodegradable, renewable, and eco-friendly nature of vegetable oil as lubricants. Vegetable oils (HOVOs) are primarily triacylglycerides, which are formed by esterification of three hydroxyl groups of the glycerol fragment with carboxyl groups of fatty acids (see Sect. 1.1 and Fig. 1.1). Vegetable oils are readily biodegradable materials due to the presences of esters [1]. Natural esters are having the capacity to degrade aerobically, with oxygen or anaerobically, without oxygen [2]. The detailed graphical mechanism is given in Fig. 2.1. There are different types of biodegradability, one type of degradation is complete

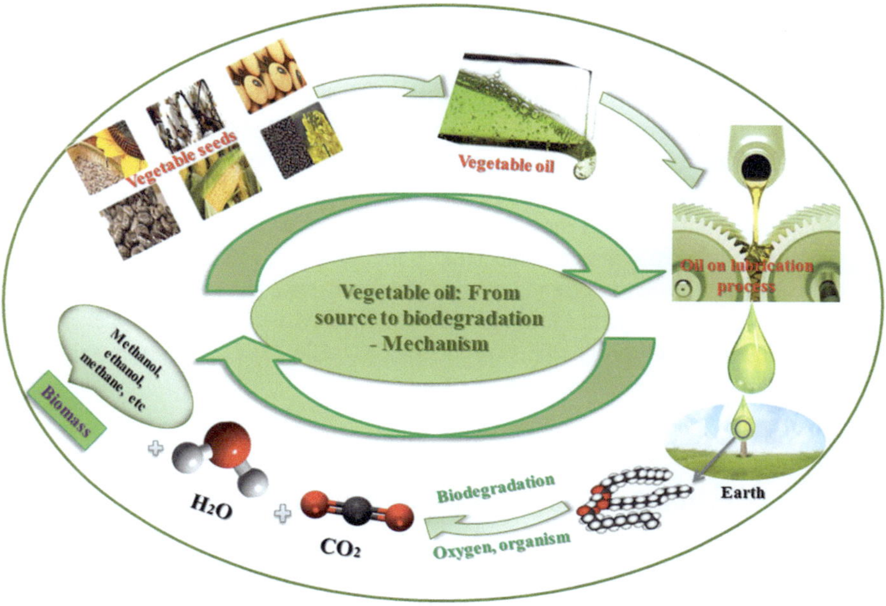

Fig. 2.1 Biodegradation mechanism of vegetable oil as lubricants

biodegradation, i.e., the lubricant has a capacity to degrade completely and returned to nature, and another one is partial biodegradability, i.e., one or more constituent of the oil is not degraded. The rate of biodegradation depends on the chemical structure of base oil or additive components. Subsequently, the biodegradation process is environment dependent, and oil may readily degrade under one condition and may remain as it is under another condition. There should be appropriate bacterial growth, adequate oxygen, and a suitable temperature for biodegradation progress. Already, the origin of food for microorganisms exists in the oil; however, an additional amount of oxygen must be present for a reasonable rate of biodegradation [3]. The degree of biodegradation is affected by the following [4]:

- Biological constituents such as amino acids, fatty acids, etc. are freely biodegradable.
- Aromatic compounds are tough to biodegrade; a benzene ring is having hydroxy, carboxylic, amine, methyl, or $-OCH_3$ compounds which somewhat biodegrade easily, but halogen, nitro, and sulfonated compounds are resistant to biodegradation.
- Linear structure carbon chain undergoes biodegradation more readily than branched compounds. [5].
- Sterically hindered bulky ester linkages decrease the biodegradability to a greater level. The biodegradation process depends on the position and degree of chain branching.

- The biodegradability products of transesterified vegetable oil decrease with acyl and alcohol chain length.

The chemical structure and properties of vegetable oil provide the best biodegradability when compared with other type of lubricants. Some measurable, specific changes of chemical and physical properties of the material occur by microorganisms in the primary biodegradation process. Therefore, this leads to the negligible transformation that modifies the physical properties of a compound, whereas leaving the molecule mostly intact. Intermediary metabolites or intermediary biodegradation may cause more toxicity than the original substrate [6]. In the other case, the complete biodegradation, the biodegradation achieved while a substance is completely utilized by microorganisms, produces CO_2, CH_4, H_2O, inorganic salts, and fresh microbial cellular components [1, 2].

2.2 Biodegradation Mechanism of Vegetable Oil

The mechanism of biodegradation is the ability of a vegetable oil molecules to be degraded biologically, i.e., by the action of biological organisms on vegetable oil. The simple mechanism of biodegradation and the life cycle of bio-lubricants are as shown in Fig. 2.1. In the case of bio-lubricants, numerous methods are available to check biodegradability nature of oil, and those are explained as follows:

2.2.1 Methods of Biodegradation Mechanism

Biodegradability is measured by some specific parameters and is considered to be revealing of the ingestion of the oil by bacteria and the formation of small molecules like CO_2, CH_3OH, H_2O, etc. [1] (see Fig. 2.1). Therefore, there are several methods which quantize the amount of minerals formed during a test period. The process which measures the loss of dissolved organic carbon for water-soluble compounds and there are gauge the uptake of oxygen by the biochemical oxygen demand (BOD).

2.2.1.1 ASTM D5864 Test Method Determines Biodegradation of Lubricant

The method determines the degree of aerobic aquatic biodegradation of lubricating oils when exposed to an inoculum under ambient conditions. The inoculum is the activated sewage sludge generated through domestic wastewater, or it may be produced from soil or natural surface waters, or any mixture of three sources. The degree of biodegradability is measured by calculating the rate of conversion of the lubricant to CO_2. By this test method, turbine oil, hydraulic fluid, or grease is classified as readily biodegradable when 60% or more of the material test carbon is converted to CO_2 in 28 days.

2.2.1.2 ASTM D6139 Test Method Determines Biodegradation of Lubricant

The test procedure determining the actual aerobic aquatic biodegradation of lubricating oil or their components using the Gledhill shake flask.

2.2.1.3 Method CEC-L-33-A-94

The available most suitable, recognized test methods employed by the lubricant industry in evaluating the rate of biodegradation of their products. This is developed by the Co-ordinating European Council (CEC); primary biodegradability is CEC-L-33-A-93 modification of some physical and chemical properties of the substance caused by the activity of microorganisms. Tables 2.1 and 2.2 showed the biodegradable criterion and biodegradable nature of different types of oils.

2.2.1.4 Method OECD 301B (CO_2 Evolution)

Organization for Economic Co-operation and Development (OECD)

Inherent biodegradability: OECD 302 would be reached under optimum conditions.

Ultimate biodegradability: OECD 301—complete consumption of the substance and conversion to carbon dioxide or methane, water, some mineral salts, and microbial cellular constituents, i.e., biomass. Tables 2.1 and 2.2 showed the % and test methods of biodegradation of different types of oils.

2.2.1.5 Method EPA 560/6-82-003

Number CG-2000, the Shake Flask Test, adapted by the US Environmental Protection Agency (EPA). These tests determine the degree and extent of aerobic aquatic biodegradation under ambient conditions. The Modified Sturm Test and Shake Flask Test also calculate the rate of transformation of the lubricant to CO_2. The % of biodegradation of all types of oils is shown in Table 2.2. Through infrared spectroscopy, the CEC test observes the vanishing of the lubricant molecules by evaluating test material at various incubation times. Laboratory test methods have been shown that the degradation quantities may vary widely among the diverse test methods indicated above [6–8].

2.3 Vegetable Oil: A Nontoxic

The lubricant and its additive toxicity have the tendency to generate the adverse biochemical effects on living organisms.

Toxicity is nothing but which is the degree to which a substance can damage an organism. The "nontoxic" claim implies that the product, material, or chemical will not cause adverse health effects, either immediately or over the long term. Importantly, environmentally friendly oil not only has biodegradability but also nontoxicity such that they are not capable of harming flora and fauna. Generally, toxicity tests are carried out to analyze the effect of a substance on the flora and fauna in the environment. The obtained results are mentioned as LD (lethal dosage), and the dosage will hinder 50% growth of the population [3]. The general rules have been followed for ecotoxicity, and materials with an LD value greater than 1000 ppm are referred as low-toxicity or nontoxic materials. Actually, ecotoxicity of substance denotes the toxicity of oil only on plants and animals [8]. Generally, toxicity of mixtures (base oil + additives) is found to be close to the arithmetic sum of individual component's toxicities. Hence, the toxicity of fully blended lubricants is depended on the base oil and its additive components. The most useful general information on the toxicological behavior of base oils can be found in CONCAWE, 1987. As per the standards laid out by CONCAWE, vegetable-based oils or lubricants are 100% nontoxic.

To evaluate the toxicity, typical principles have been followed, and they are:

- Aquatic oral and dermal toxicity—LD50 < 2000 ppm
- Eye irritation, corneal opacity—2000 ppm or more
- Conjunctival redness (24–72 h)—2000 ppm or more
- Skin irritation anathema (redness)—2000 ppm or more

In order to improve environmentally suitable lubricants, toxicological criteria must be considered. Different countries have adopted to follow several test methods to analyze the criteria of toxicity of lubricants. The primary factor is to protect the life in various areas such as in water (aquatic area) and nonaquatic environments. There are several standard test procedures or methods to test the toxicity of vegetable oils, and they are as follows:

Bacterial toxicity by *DIN 38412–8* method—this method determines inhibition of cell multiplication of effective concentration (EC) (EC_{10} and EC_{50} values). The *Pseudomonas* bacteria utilized for analysis are established in wastewater and soil. Bacterial testing, according to the ISO 8192 method, concludes acute toxicity through the inhibition of oxygen consumption; the results are expressed as EC_{50} values [1]. The algal toxicity test is according to *DIN 38412–9* which is a further test for aquatic systems (measurement of chlorophyll fluorescence and determination of EC_{10} and EC_{50} values). The "Daphnia test" is according to *DIN 38412–11* or OECD guideline 202, which is a method for small living organisms and one of the most important test procedures in German legislation concerning the aquatic environment. Fish toxicity, according to *DIN 38412–15*, is performed on the Goldorfe

Table 2.3 Toxicity of lubricants—test methods [1, 2]

International methods	Bio-organism	Parameters
OECD 201/DIN 38412 (IX)	Algae	EC10, EC50
OECD 202/DIN 38412 (IX)	Daphne	EC10, EC50
OECD 203	Fish	LC10, LC50, LC100
OECD 209	Sludge	LC10, LC50, LC100
ISO 8192/DIN 38412 (VII)	Bacteria	EC10, EC50

Table 2.4 Aquatic toxicity limits based on EC50 values [1, 2]

Classification	*EC50 mg/l*		
Relatively harmless	>1000		
Practically nontoxic	>100–1000		
Slightly toxic	>1–100		
Moderately toxic	>1–10		
Highly toxic	<1		
German classification of fluids based on WGH numbers			
Materials	*WEN*	*WGK number*	*Classification*
Vegetable oils and esters	0–1.9	0	Not hazardous to water
Mineral lubricant base oil and white oils	2–3.9	1	Slightly hazardous to water
Additive treated lube oils and industrial oils	4–5.9	2	Slightly hazardous to water
Additive treated soluble oil, water miscible coolants	>6	3	Highly dangerous to water

(*Leuciscus idus*), and the results are expressed as LC_0, LC_{50}, and LC_{100} values. The lethal dose (LD_{50}) is an important measure of toxicity. Fish toxicity, according to OECD guideline 204, is measured as part of the German eco-label "Blue Angel." Potential pollution of the nonaquatic area, e.g., soil and plants, is evaluated with the plant growth test according to OECD guideline 208 (for wheat, cress, and rapeseed).

A list of such methods is given in Table 2.1, and the criteria for aquatic toxicity levels based on EC50 values are reported in Tables 2.3 and 2.4.

2.4 Environmental Friendly Lubricating Oil

A lubricant at the best can only be environmentally neutral. The lubricating oils can be accepted as environmentally friendly when it has met the following basic criteria [8, 9].

At the manufacture or fabricating stage, the lubricant must be environmentally friendly (green) and neutral, i.e., it consumes less energy, produces no waste materials, and creates no emission:

(a) The lubricant would not create any disposal harms, i.e., should be easily biodegradable.
(b) The lubricant would be physiologically harmless, i.e., no toxicity and noncancerous.
(c) The lubricants must have quick biodegradability and would not create any toxic decay products.
(d) The base lubricating oil has to be refined from a renewable source, if possible; if so, there is no depletion of natural resources; there is no addition to the greenhouse effect.
(e) Nontoxic substances; nonbioaccumulative potential.
(f) The lubricant would be ecotoxicologically acceptable, i.e., non-water contamination, non-water miscibility, and low density than H_2O.

2.4.1 Environmental Friendly Lubricants: Vegetable Oil as Bio-lubricants

Presently, many of the oil industries have been developing "environmentally friendly" oils or bio-lubricants, because of increasing concern related to environmental impact and associated costs of lubricants. On the other hand, there is no universal description used to determine the environmental friendliness. Actually, bio-lubricants are nothing, but additive package blended with base oils to improve rapid biodegradation character and lower ecotoxicity as primary target or objective. Even though, there is no worldwide general contract on the origin and chemical composition of such bio-lubricants. Bio-lubricants are often, but not necessarily, based on vegetable oils. During the fabrication of a biodegradable and low-toxicity lubricant, it is of responsibility to select the additive which must biodegrade and is less toxic in nature [1]. Bio-lubricants exhibit a neutral CO balance and readily decompose in nature. At present marketed bio-lubricants, i.e., vegetable oil, are an environmentally sensible substitute with performance features and quality compared to mineral-based competitors. Due to a high boiling temperature range of natural esters, bio-lubricants show very less emission and are entirely free from aromatic compounds, and over 90% are biodegradable in nature and non-water polluting [4, 5, 8]. Likewise, the oil haze and an oil vapor reduction lead to less inhalation of oil mist into the lung. Bio-lubricants have beautiful human skin compatibility, less dermatological problems, as well as high cleanliness in the working environments. The high wetting affinity of polar esters leads to friction drop with at least equal and often greater instrument life. The high viscosity index (VI) of vegetable oil can be benefited when fabricating lubricants for use wide range of temperature (see Table 7.5). It can be useful as less viscosity grade for same applications combined with an easier heat transfer. It has high safety in case of flash and fire-related properties due to its high flash points at the same viscosities leading to low cost on account of less maintenance, manpower, and storage and disposal costs. The relative rating of vegetable oil properties with mineral and synthetic oil is discussed in Table 2.5.

Table 2.5 Rating of vegetable oil properties compared with mineral and synthetic oil [8–11]

Oil properties	Mineral oil	Polyalphaolefins	Polyalkylene glycols	Dicarboxylic acid esters	Neopentyl polyesters	Vegetable oil, e.g., rapeseed oil	References
Viscosity index	D	B	B	B	B	B	Bartz (1998), Bartz (2006), Nagendramma and Kaul (2012) and Rudnick (2006) [8–11]
Pour point	E	A	C	A	B	C	
Thermal stability	D	D	C	C	B	D	
Hydraulic stability	A	A	C	C	D	E	
Oxidative stability	D	B	C	B/C	B	E	
Flash point	E	E	D	D	D	E	
Corrosion protection	A	A	C	D	D	A	
Toxicity	C	A	C	C	C	A	
Biodegradability	D	C/D	A/B	A/B	A/B	A	
Lubricating property	C	C	B	B	B	A	
Miscibility with mineral oil	–	A	E	B	B	A	
Volatility	D	B	C	A	A	C	
Seal material compatibility	C	B	C	D	D	D	
Solubility of additives	A	B	D	B	B	C	
Cost relation with mineral oil	–	C/E	C/E	B/E	B/E	B/C	

Assessment: *A* excellent, *B* very good, *C* good, *D* moderate, *E* poor

2.4.2 Suitable Structure and Lubrication Properties of Vegetable Oil: A Environmental Friendly Lubricants

Majority of vegetable oils contain mainly triacylglycerides (natural esters). Vegetable oils, i.e., triacylglyceride structure, contain predominantly long chains of hydrocarbons (carbon and hydrogen) with the terminal functional group having carbon, hydrogen, and oxygen atoms. These differentiate them from the mineral oil chemical structure which consists of chains and rings of carbon and hydrogen. It has one of the most important properties entirely different from mineral oil, i.e., the carbons may or may not be saturated with hydrogen. Because of the presence of triglyceride units, vegetable oils are more polar than petroleum-based mineral oils. As a result, triglycerides have a higher affinity to metal surfaces [12] (see Sect. 1.3 and Fig. 1.3). Triacylglycerides, also named as triglycerides, are glycerol molecules with three long-chain fatty acids attached at each hydroxyl group via ester linkages (see Fig. 1.1). The structures of natural esters have different functional groups; they are double bonds, allylic carbons, and ester groups, which are potential places for chemical reaction or modification. The triglyceride structure of vegetable oils offers quality properties of a lubricant. Mainly, long polar fatty acid chains of triglycerides provide enough strength for boundary lubricant films which interact strongly with metallic surfaces, hence reducing both the friction and wear. The strong intermolecular interactions are flexible to changes in temperature giving a high viscosity index. Vegetable oils have several important properties that are required in a lubricant; they are low volatility, high viscosity index, high lubricity, and advanced properties like low toxicity and high biodegradability that can be compared to mineral oil [9]. Relative properties of vegetable oil compared with other oils are given in Table 2.6. Generally, fatty acids are present in natural vegetable oils which differ in the number of repeating hydrocarbons and number of unsaturated bonds. The unsaturated fatty acid composition is determined by the degree and position of carbon-carbon double bonds. The lubricating properties of the vegetable oil depend on the fatty acid structure. The vegetable oil properties vary with the fatty acid units. The long-chain carbon is alleged together with one, two, or three double bonds: oleic, linoleic, and linolenic fatty acid components, respectively (see Table 1.5). Most of the vegetable oils contain at least 4 and occasionally as many as 12 different fatty acids [11]. The effect of unsaturation, chain length, and branching of the lubrication properties of the oil is given in Table 2.7. Most vegetable oils of triglyceride molecule have 18 carbon chains. There was an old principle, i.e., the longer the chain length, the better the lubricant performance, but the concept may not be true in numerous cases. For example, erucic acid contains 22 carbons that are not as good as lubricating oil as canola have 18 carbons. Canola also has less amount of palmitic and stearic acids in its oil [18].

The rate of flow of vegetable oil is mostly decided by chain length, efficiency of molecular packing, intermolecular interactions, and molecular weight. Saturates have too a high level of molecular packing, which accelerates interlocking of the

Table 2.6 Important lubricating properties of vegetable oil compared with other oils

Oil nature	Viscosity at 40 °C, (cSt)	Viscosity at 100 °C, (cSt)	Viscosity index ISO 2909	Pour point, (°C)	Flash point, (°C)	Air release (mins) ISO 9120	Foaming (ml/min) ISO 6247	Demulsibility (minutes) ISO 6614	References
Coconut oil	27.7	6.1	175	–	–	–	–	–	Nagendramma and Kaul (2012), Rudnick (2006), Talkit et al. (2012), Soni and Agarwal (2014), Suarez et al. (2010), Syahrullailet al. (2011), Syahrullail et al. (2013), Erhan and Asadauskas (2000) and Quinchia et al. (2014) [10–18]
350 neutral mineral oil	65.6	8.4	97	−18	252	–	–	–	
Low erucic rapeseed oil	36.2	8.2	211	−18	346	–	–	–	
High oleic sunflower oil	39.9	8.6	206	−12	252	–	–	–	
Conventional soybean oil	28.9	7.6	246	−9	325	–	–	–	
Palm oil	39.7	8.2	188	N/A	N/A	–	–	–	
Mineral oil	40	–	100	–	–	5	10/0	10	
							40/0		
							20/0		
Rapeseed oil	38	–	215	–	–	2	20/0	20	
							50/0		
							40/0		
Unsaturated ester	40	–	185	–	–	1	10/0	20	
							10/0		
							10/0		
Saturated ester	40	–	145	–	–	1	0/0	10	
							0/0		
							0/0		

Table 2.7 Effect of fatty acid unsaturation, chain length, and branching on the lubrication properties of base oil [16–23]

	Lubricity	VI	Low-temperature fluidity	Oxidation stability					Volatility	References
Chain length	++	+	+	–					+	Syahrullail et al. (2013), Erhan & Asadauskas (2000), Quinchia et al. (2014), Sharma et al. (2006), Gawrilow (2004), Willing (2001), Zaher & Nomany (1988) and Hwang et al. (2003) [16–23]
Chain branch	–	–	–	+					–	
Unsaturation	–	+/–	+		*Steric*	*Oleic*	*Linoleic*	*Linolenic*	+/–	
				Double bond	18:0	18:1	18:2	18:3		
				Relative oxidation stability	1	10	100	200		

++ very positive effect, + positive effect, +/– without effect, – very negative effect

needlelike triacylglycerol crystals as temperature reduced. The vegetable oils and their double bonds influence low-temperature behavior. Most of the vegetable oils are readily available and at low cost and are not suitable for lubrication due to their high saturates or polyunsaturated fatty acid content. The vegetable oil has monounsaturated fatty acid which presents most favorable oxidative stability and low-temperature properties. As a consequence, vegetable oils that have high stability and low pour points can be produced by converting all the fatty acids into a monounsaturated fatty acid. Thus, base oil for lubrication application must have a balance of fatty acids, preferably a high level of monounsaturated, minimal polyunsaturated, and ideally no saturates at all for cold climates [20].

2.5 Literature Survey of Vegetable Oil as a Lubricant

The world's largest sources of vegetable oils are plants such as soybean, corn, linseed, cottonseed, and peanuts. But other sources are oil-bearing perennials such as the palm, olive, or coconut [24]. The worldwide production of soybean oil ranks first (29%) among vegetable oils and represents the cheapest and readily available source of plant oil in the world. Other types of vegetable oils include canola, mainly formed in Canada, rapeseed in Europe, and castor beans, olives, sesame, corn, sunflower, peanuts, cottonseed, coconut, palm fruits, palm kernels, linseed, safflower, and so on. Naturally occurring plant oils and derivatives of fatty acids are the most important renewable feedstock. Those renewable plant oils are processed in the chemical industry and are prepared as bio-based functional polymers and polymeric materials [25–30].

2.5.1 Vegetable Oil as a Base Stock for Lubricants

Vegetable oils have a plenty of essential potentials which provide them benefits over mineral fluids as the base stock in lubricants, and the detailed differentiations (advantage and disadvantages) are shown in Table 2.8. From the literature survey, vegetable oils are obtained from a renewable resource which keeps away from the upstream pollution associated with petroleum removal and purifying regarding usage [12]. Figure 2.2 displayed the difference in the source of lubricant fluids and its application. From a worker's safety viewpoint, vegetable oil-based bio-lubricants are smarter than petroleum counterparts due to their comparative low toxicity, high flash point, and low volatile organic compound (VOC) emissions. The performance limitations of vegetable-based lubricants stem from inherent properties of base stocks rather than additive composition. Usually, base stocks are having the predefined properties of low volatility, ideal cleanliness, high biodegradability, high solvency for lubricant additives, miscibility with another type of fluids, negligible effects on seals, elastomer compatibility, density, or heat conductivity. To determine

Table 2.8 Advantages and disadvantages of vegetable oil

Advantages	Disadvantages	References
Very high viscosity indexes	Poor oxidative stability	Soni and Agarwal (2014), Bartz (1998), Bartz (2006), Fox and Nagendramma and Kaul (2012), Goyan et al. (1998), Samarth and Mahanwar (2015), Jumat et al. (2010) and Gawrilow (2003), [13, 8–10, 31–33]
Good thermal stability	Poor hydrolytic stability	
Low volatility	Poor low-temperature characteristics	
High flash points	Poor response to pour point depressants	
Good miscibility with other lubricant base fluids and solvents	Tendency to clog the filter	
Good biodegradability and low toxicity	A narrow range of viscosities	
Good solvents		
Load carrying capacity		
Anti-wear properties		
Protection against rust and copper corrosion		
Good filterability		
Good additive compatibility		
Better fire resistance		
Poor resistance to foaming		
Approximate relative costs of various base fluids are:		
Refined mineral oils, 1		
Vegetable oils, 1.5–2		
Synthetic esters, 4–12		
Nontoxicity		
Biodegradability		
Resource renewable		
Low emission of CO and hydrocarbons		
Less CO_2 production		
Affordable application cost		
Good lubricity—polar nature of the ester linkage imparts good lubricity but also competes with the surface-active additives		

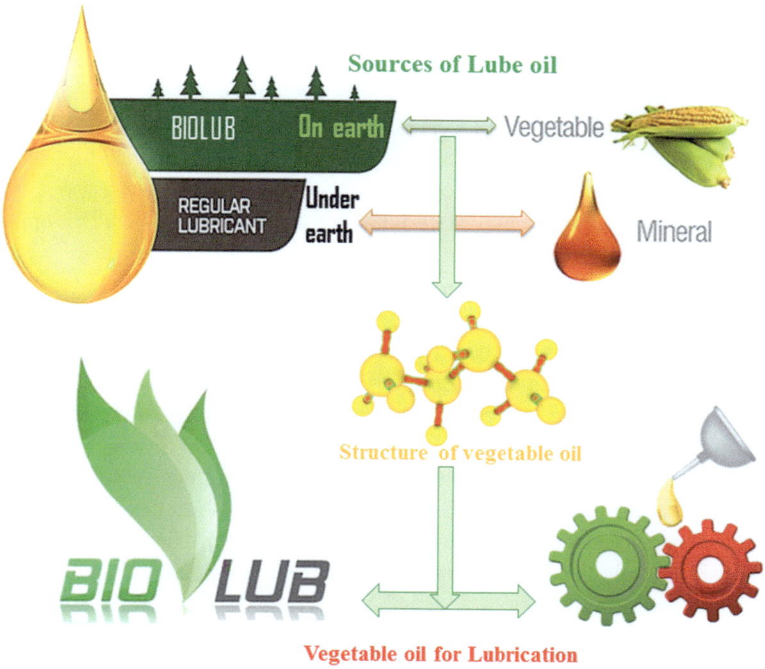

Fig. 2.2 Graphical representation of sources of fluids (vegetable and mineral oil) and its application

oxidative stability, deposit foaming tendencies, low-temperature solidification, viscometric properties, and hydrolytic stability, the base stocks play an important role. On the other hand, the important lubricating properties such as lubricity, antiwear protection, load carrying capacity, corrosion and rust prevention, acidity, ash content, color, foaming, demulsibility nature, water rejection, and some others are mostly dependent on the additives and contaminants. Hence, while choosing oil, in which its suitability as a lubricant should be considered, first of all, the base stock-dependent parameters are evaluated. In addition to the degradation capacity of lubricants, some of the properties to be given importance are particle count (cleanliness), compatibility with other petroleum products and uniformity during long-term storage, moisture content and acidic nature, viscosity, viscosity index, pour point, cloud point, volatility, oxidative stability, elastomer compatibility, and other properties, depending on proposed application [17]. However, vegetable oils are facing major issues, i.e., little resistance to oxidative degradation and poor low-temperature properties, and it had to be taken care of. The researchers and scientists had to find out, and it has been recognized that methylene-interrupted polyunsaturation is the key factor causing the lower oxidative stability of vegetable oils [22, 23].

2.6 Status of Vegetable Oil

In the Indian context, enormous nonedible oils remain not yet used, which would be utilized as a potential source for vegetable oil-based lubricants [34]. Most of the nonedible oilseeds are growing across the railway lines, wastelands, highways, and fencing of various types. When we consider seeds like castor, *Jatropha*, karanja, mahua, tamanu, etc., the product price will further be reduced [35]. Presently, four primary vegetable oils lead the industry accounting for about 82–85% of worldwide vegetable oil manufacture. Soybean oil ranges 31–35%, palm oil ranges 28–30%, rapeseed oil ranges 14–15%, and sunflower oil ranges 8–10% of worldwide production [12]. India is the biggest exporter of castor oil, and it corners around 70% of the global castor oil trade. India provided the preference to castor oil; it is constantly in demand predominantly in the industrial sector, which is estimated at 220000 tons per annum. India consumes nearly 25% of its own production of which one-fourth is used in the soap industry, while lubricant industry consumes one-fifth. Paint and allied industry consumes nearly 35% of the domestic consumption. The castor oil is one of the most expensive vegetable oils in the international market. Despite India's dominant situation in the international castor in the export market, the Indian castor produce does not drive the prices. It distributes around 0.2–0.24 mn tons of commercial castor oil, 50,000–60,000 tons of castor seed extractions, and 15,000–20,000 tons of castor seed. India supplies 70% of the world's requirements of castor oil [31].

2.6.1 Action Mechanism of Vegetable Oil in Lubricating Process

Natural ester of vegetable oils can act as an antiwear additives and friction modifiers, due to strong interactions with the lubricated metal surfaces. Its amphiphilic (having both hydrophilic and hydrophobic parts) nature of plant oil gives them a good film association on metal surfaces, because of hydrophobic long fatty acid chains and the presence of hydrophilic polar groups in its structure [36, 37] (see Fig. 1.3). For this adaptable reason, vegetable oil has the peculiarity of being active as both boundary and hydrodynamic lubricants [36–39]. To completely understand the tribological properties of plant oils, it is essential to comprehend the consequence of the unevenness in fatty acid ratio on their lubricating properties, film thickness formed, friction, and wear [40]. Oils with more polar groups (such as esters and carboxylic acids) have more reactive sites to adsorb with metal surfaces to supply boundary lubrication effects [41]. The lubricating film with strong bonding to the metal surface and enough cohesive interaction between lubricant reactive units can reduce the friction and the quantity of wear. Hence, to keep a low friction and wear, the lubricating film has to withstand extreme pressure and temperature and shear degradation and maintain excellent boundary oiling properties via active physical and chemical adsorption with the metal (see Sect. 1.3 and Fig. 1.3). Vegetable oils are known to have superior essential merits like good biodegradability and lubricity, enhanced flash and fire points, lower toxicity, much higher viscosity, and viscosity index. The lubricity of vegetable oils is attributed to their ability to form a tenacious monolayer of the polar head stick to the metallic surfaces and the long carbon chains orienting in near normal directions to the surface. Biodegradable vegetable oils are perceived as base oil substitutes for mineral oils in lubricants mainly because of their superior environmental characteristics. These triacylglycerol molecules in vegetable oils have oriented themselves with the polar end at the solid surface making them a packed closed, mono- or multimolecular layer resulting in a surface film that offers desirable potentials in a lubricant [42]. It not only has those properties, but it also has other advantages like very low evaporation characteristics due to the high molecular weight of the triglyceride molecule.

2.7 Conclusion

The conspectus of the chapter describes the importance of eco-friendly, biodegradable, renewable, and profitable green lubricants as a base stock which can replace petroleum base stocks in the next generation of industrial lubricants. In recent scenario, the vegetable oils are naturally suitable to be used as lubricant base oils. Current literature survey reports the development of novel additives and chemical modification processes of vegetable oils to improve the oxidative and hydrolytic stabilities to make them suitable as base stocks that are performing better than the mineral and synthetic oils. Plant oils, i.e., vegetable oils, are considered to be substitutions to mineral oils as base stocks for lubricants because of developing environmental concerns. Lubricants refined from vegetable oil provide important environmental benefits with respect to resource renewability, biodegradability, low toxicity, and being cheaper than synthetic oils and offer satisfactory performance in a wide range of applications. However, vegetable oil-based bio-lubricants are an important part of new strategies, policies, and subsidies, which aid in the reduction of the dependence on mineral oil and other nonrenewable sources. It gives detailed literature survey of modified vegetable oils that have suitable chemical skeleton structure and appropriate properties pertaining to physical, chemical, physicochemical, and tribological to perform as viable eco-friendly, biodegradable, renewable green lubricants.

References

1. Anand ON, Kumar CV (2006) Vegetable oil derivatives: environment-friendly lubricants and fuels. J Synth Lubr 23:91–107
2. (a) Steve B (2002) Green lubricants. Environmental benefits and impacts of lubrication. Green Chem 4:293–307. (b) Liew Yun Hsien W (2015) Towards green lubrication in machining, Springer briefs in green chemistry for sustainability. Springer, Malaysia

3. Goyan RL, Melley RE, Wissner PA, Ong WC (1998) Biodegradable lubricants. Lubr Eng 54:10–17
4. Kitamura N (1993) Biodegradable lubricants. Jpn J Tribol 38:639
5. Erhan SM, Kleiman R (1997) Biodegradation of estolides from monounsaturated fatty acids. J Am Oil Chem Soc 74:605
6. www.wiserenewables.com (2006) Wise Solutions–Renewable Lubricants–Biodegradability Primer
7. Aluyor EO, Obahiagbon KO, Orijesu M (2009) Biodegradation of vegetable oils: a review. Afri J Biotechnol 8:915–920
8. Bartz WJ (1998) Lubricants and the environment. Tribol Int 31:35–47
9. Bartz WJ (2006) Ecotribology: environmentally acceptable Tribological practices. Tribol Int 39:728–733
10. Nagendramma P, Kaul S (2012) Development of ecofriendly/biodegradable lubricants: an overview. Renew Sust Energ Rev 16:764–774
11. Rudnick L. R (2006) Synthetics, mineral oils, and bio-based lubricants: chemistry and technology, CRC/Taylor & Francis Group, New York
12. Talkit KM, Mahajan DT, Masand VH (2012) Study on physicochemical properties of vegetable oils and their blends use as possible ecological lubricant. J Chem Pharm Res 4:5139–5144
13. Soni S, Agarwal M (2014) Lubricants from renewable energy sources – a review. Green Chem Lett Rev 7:359–382
14. Suarez AN, Grahn M, Pasaribu R, Larsson R (2010) The influence of base oil polarity on the tribological performance of zinc dialkyl dithiophosphate additives. Tribol Int 43:2268–2278
15. Syahrullail S, Zubil BM, Azwadi CSN, Ridzuan MJM (2011) Experimental evaluation of palm oil as lubricant in cold forward extrusion process. Int J Mech Sci 53:549–555
16. Syahrullail S, Kamitani S, Shakirin A (2013) Performance of vegetable oil as lubricant in extreme pressure condition. Procedia Eng 68:172–177
17. Erhan SZ, Asadauskas S (2000) Lubricant base stocks from vegetable oils. Ind Crop Prod 11:277–282
18. Quinchia LA, Reddyhoff T, Delgado MA, Gallegos C, Spikes HA (2014) Tribological studies of potential vegetable oil-based lubricants containing environmentally friendly viscosity modifiers. Tribol Int 69:110–117
19. Sharma BK, Adhvaryua A, Liu Z, Erhan SZ (2006) Chemical modification of vegetable oils for lubricant applications. J Am Oil Chem Soc 83:129–136
20. Gawrilow I (2004) Vegetable oil usage in lubricants. Oleochem Inform 15:702–705
21. Willing A (2001) Lubricants based on renewable resources an environmentally compatible alternative to mineral oil products. Chemosphere 43:89–98
22. Zaher FA, Nomany HM (1988) Vegetable oils and lubricants. Grasas Aceites (Seville) 39:235–238
23. Hwang HS, Adhvaryu A, Erhan SZ (2003) Preparation and properties of lubricant base stocks from epoxidized soybean oil and 2-ethylhexanol. J Am Oil Chem Soc 80:811–815
24. Hui YH (1995) Bailey's industrial oil and Fats products, edible oil and fat products: general application, vol 1, 5th edn. Wiley, Blackwell, pp 19–44
25. Guner FS, Yagci Y, Erciyes AT (2006) Polymers from triglyceride oils. Prog Polym Sci 31:633–670
26. Habib F, Bajpai M (2011) Synthesis and characterization of Acrylated Epoxidized soybean oil for UV-cured coatings. Chem Chem Technol 5:317–326
27. Ronda JC, Lligadas G, Galià M, Cádiz V (2011) Vegetable oils as platform Chemicals for Polymer Synthesis. Eur J Lipid Sci Technol 113:46–58
28. Montero de Espinosa L, Meier MAR (2011) Plant oils: the perfect renewable resource for polymer science. Eur Polym J 47:837–852
29. Gnanasekaran D (2014) POSS Nanocomposite membranes: Synthesis and gas separation studies. Scholars press, Saarbrücken

30. Gnanasekaran D, Reddy BSR (2013) Cost effective poly (urethane-imide)-POSS membrane for environmental and energy-related processes. Clean Techn Environ Policy 15:383–389
31. Samarth NB, Mahanwar PA (2015) Modified vegetable oil based additives as a future polymeric material-review. Open J Organic Polym Mater 5:1–22
32. Jumat S, Nadia S, Emad Y (2010) Biolubricants: raw materials, chemical modifications and environmental benefits. Eur J Lipid Sci Technol 112:519–530
33. Gawrilow I (2003) Palm oil usage in lubricants. Paper presented on 3rd Global Oils and Fats Business Forum USA; Interfacing with the Global Oils and Fats Business
34. (a) Joseph PV (2007) Study of some non-edible vegetable oils of Indian origin for lubricant application. J Synth Lubr 24:181–197. (b) Chauhan PS, Chhibber VK (2013) Non-Edible oil as a source of bio-lubricant for industrial applications: a review. Inter J Eng Sci Innov Technol 2(1):299–306
35. Sahoo D, Jena KS, Rout PK, Rao YR (2002) Characteristics of Simarouba glauca seeds and fat of Orissa origin. J Food Sci Technol 39:552–555
36. Adhvaryu A, Erhan SZ (2004) Tribological studies of thermally and chemically modified vegetable oils for use as environmentally friendly lubricants. Wear 257:359–367
37. Biresaw G (2006) Elastohydrodynamic properties of seed oils. J Am Oil Chem Soc 83:559–566
38. Maleque M, Masjuki H, Sapuan S (2003) Vegetable-based biodegradable lubricating oil additives. Ind Lubr Tribol 55:137–143
39. Biresaw G, Bantchev G (2008) Effect of chemical structure on film-forming properties of seed oils. J Synth Lubr 25:159–183
40. Stachowiak GW, Batchelor AW (2005) Engineering tribology, 3rd edn. Elsevier, Oxford
41. Rudnick LR (2009) Additives for industrial lubricant applications. Lubricant additives chemistry and applications. CRC Press Taylor & Francis Group, Boca Raton
42. Bilal S, Mohammed-Dabo IA, Nuhu M, Kasim SA, Almustapha IH, Yamusa YA (2013) Production of biolubricant from Jatropha curcas seed oil. J Chem Eng Mater Sci 4:72–79

Chapter 3
Vegetable Oil as a Multifunctional and Multipurpose Green Lubricant Additive

Abstract It covers the different types of biodegradable triglycerides of vegetable oil (soybean oil, sunflower oil, jojoba oil, natural garlic oil, etc.) which functioned as multifunctional and multipurpose lubricant additives. In lubricants, each conventional additive improves particularly only one property, i.e., pour point, antiwear, viscosity improver, etc. But in the case of recent studies on additives, natural esters of vegetable oil which are used as multipurpose green additives are discussed in detail.

Keywords Multifunctional additives • Multipurpose lubricant additives • Pour point depressant • Viscosity index improver

3.1 Introduction

Studies and research activities of mineral oil over the past six decades have improved the oil properties and expanded their applications in all sorts of rotating and sliding parts of vehicles and machinery. Especially, the base oils cannot fulfill all the requirements of modern engines. Hence, a large number of functional additives are added to the base oil to enhance the characteristic properties already present or to impart some new additional properties. Generally, lubricant additives are polymeric in nature; they enhanced service performance of the virgin lubricating oil. The effectiveness of polymeric additives mainly depends on their morphology, which is directly linked to the procedure of polymerization. Usually, the linear polymer structures with narrow molecular weight distribution are more effective as additives in the paraffinic base oil. In general, additive molecules have long oil-soluble, non-polar (hydrocarbon) tails and smaller polar (hydrophilic) head groups. The chemical structure and skeleton of conventional additive and triglyceride (natural ester) as an additive are shown in Fig. 3.1. Therefore, additives have a tendency to be present in the colloidal form as inverse micelles due to the two parts of the molecule have differed in the nature of solubility in the fluids. Their amount varies from 1 to 30% or more depending on the applications.

D. Gnanasekaran, V.P. Chavidi, *Vegetable Oil based Bio-lubricants and Transformer Fluids*, Materials Forming, Machining and Tribology,
DOI 10.1007/978-981-10-4870-8_3

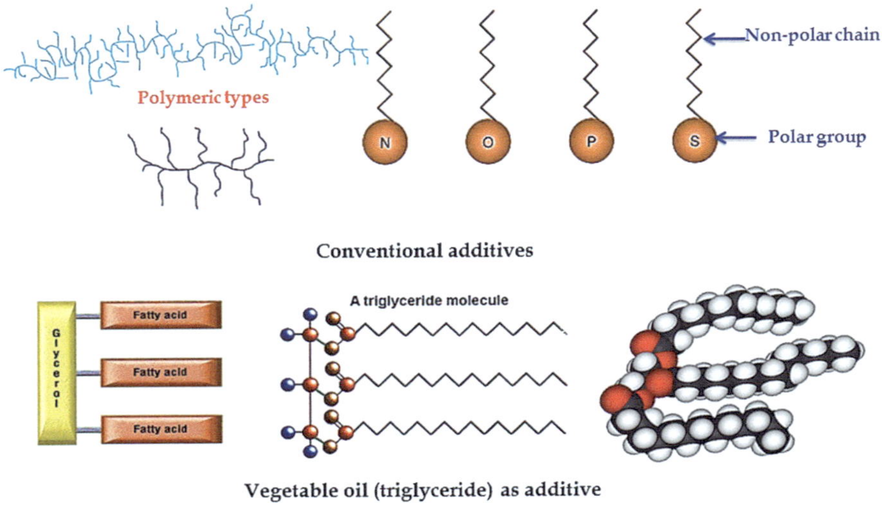

Fig. 3.1 The basic chemical structure of conventional additives and vegetable oil as additives

Lubricant additives are used to improve pour point (PP) of the base oils by dissolving wax crystal deposits at low temperature, increase viscosity index, minimize wear, carry away contaminants and debris, etc. Pour point depressants (PPDs), viscosity index improvers (VIIs), antiwear and extreme pressure additives, dispersants/detergents, antioxidants, etc. are an example of additives generally blended with base stocks [1–9] (Fig. 3.2). The basic formulation and systematic interactions of the additive with base oil are displayed in Fig. 3.2. Since multifunctional additives induce more than one of the above performances, hence, research throughout the world is increasingly directed toward producing such type of additives. Acrylate types of polymer additives have been widely used for a long time [8]. The usual acrylate-based additives are extremely harmful to the environment. The commercial synthetic acrylate additives are nonbiodegradable, and their widespread use has raised many environmental concerns. Due to increasing environmental pollution, the direction toward the development of an environmental kind of green polymeric additive in the lubricant chemistry is increasing day by day (see Sect. 1.3 and Figs. 1.4 and 1.5).

Vegetable oils have many promising natural properties including good lubricity, good resistance to shear, a high flash point, and a high viscosity index and low evaporative loss over the mineral and synthetic oils [10–12]. They are primarily triglycerides, that is, tri-esters of long-chain fatty acids (both saturated and unsaturated) combined with glycerol (see Figs. 1.1 and 3.1). But, they have some limitations such as poor oxidative and hydrolytic stability, high temperature-sensitive tribological behavior, poor cold flow properties, and gumming effect, which are considered in their application in industrial lubricants [10, 13]. However, the disadvantages can be minimized by means of chemical derivatization [14] of the

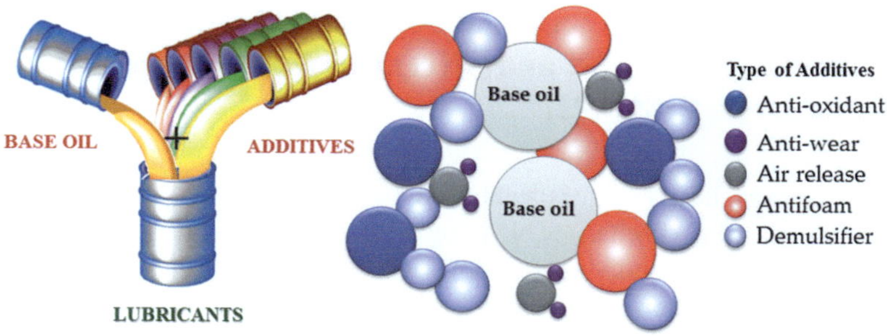

Fig. 3.2 The lubricant formulation and interaction between additive molecules and base oil

olefin groups of the vegetable oils, which lead to much better performance as well as enhanced thermal and oxidative stabilities.

Organic linear chain compounds with polar end groups, like oleic acid amide, are extensively used as friction-modifying additives in lubricants. These compounds are adsorbed from hydrocarbon solutions on a metal surface. It is generally assumed that these adsorption layers are responsible for the improved lubricating properties of oils containing these substances. The long-chain fatty acids and alcohols form monomolecular layers on metal surfaces. The molecules are adsorbed with the major axis perpendicular to the surface (see Fig. 1.3).

Research has been conducted on natural ester-based metal-working lubricant additives (organic fatty acids) on the grinding performance of the various types of carboxylic acids. The fatty acid has shown the best grinding performance of all oiliness agents, friction modifier, and EP agents. There have been many investigations concerning the chain length of the carbon atom of fatty acids on the lubricating properties (see Table 2.7). There has been a growing awareness in the use of vegetable oils, which have an excellent biodegradability and also improved chemical stability with the antiwear additive formulation. The polar additive binary system showed a lower lubrication performance compared to the individual fatty acid additive. There exist numerous examples where derivatives of vegetable oils (like soybean oil, sunflower oil, etc.) have been used as additives as well as base stocks in the formulation of eco-friendly lubricant [15–19]. Li et al. [19], in their work, have shown the application of natural garlic oil as a high-performance and environmentally friendly extreme pressure additive in lubricating oils. Biresaw et al. [20] have reported the application of biobased polyesters as an extreme pressure additive in the mineral oil. Thus, there exists much research to develop environmentally benign additives based on vegetable oils, and due to ease of availability, relatively lower cost, and better performance, considerable attention has been given to research on vegetable oil especially on lubricants and additives from soybean oil.

3.2 Homo- and Copolymers of Soybean Oil with Methyl Acrylate, 1-Decene, and Styrene as Multipurpose Additives (PPD, VII)

Gobinda and Pranab [21] synthesized lube oil additives from vegetable oil and evaluated their additive performances blending with various oils. The authors collected mineral base oils SN70 and SN150, from IOCL, West Bengal, India. They prepared diverse additives as a homopolymer of SBO (S-1) and copolymers of methyl acrylate (10%) (S-2), methyl acrylate (5%) (S-3), 1-decene (10%) (S-4), and 1-decene (5%) (S-5), respectively. In their investigation, a homopolymer of soybean oil (SBO) and its copolymers with methyl acrylate (MA), 1-decene, and styrene in two different concentrations were prepared, characterized, and compared to the performance of multifunctional additives. In those investigations, homopolymers of soybean oil and copolymers with methyl acrylate, 1-decene, and styrene were blended by a thermal technique with azobisisobutyronitrile as a radical initiator. The authors have checked the performance evaluations of the synthesized polymers as pour point depressant (PPD), viscosity index improver (VII), and antiwear in different mineral oils as per suitable ASTM test methods. Biodegradability of the prepared additives was tested against fungal pathogens and microorganisms by the disk diffusion method and soil burial test method, respectively. Table 3.1 indicated the VI and PP of differently formulated lubricants. The copolymers of soybean oil with methyl acrylate (S-2 and S-3) have very good PPD performance compared to others. The enhancement in the PPD performance by the addition of additives is due to disorder of the wax crystal chemical structure, which is formed due to the deposition of the paraffinic compounds present in mineral oil at lower temperatures (see Figs. 4.1 and 4.6). The obtained experimental confirmation shows that pour point depressing ability depends on the polarity of side chains of the additive [22]. The soy-acrylate copolymers, due to their more polar nature, act better against the creation of wax crystals and therefore showed higher pour points that are comparable with available polymethacrylates [23]. The results revealed that all the homo- and copolymers had shown excellent multipurpose acts as additives in lubricants. The viscosity index, antiwear, and pour point of the base fluids were improved considerably by the addition of these additives. The thermo-oxidative stability of soybean oil is increased by the polymerization reaction. The copolymers of SBO with 1-decene and styrene achieve better viscosity index value. The other copolymers of SBO with methyl acrylate and 1-decene behave as superior antiwear additives. The copolymer of soy-methyl acrylate has better PPD performance. Moreover, due to its being biodegradable, prepared homo- and copolymer additives are eco-friendly. Hence, the author [21] concluded the work would augment the lubricant technology to produce a cost-effective as well as eco-friendly oil composition.

Table 3.1 Viscosity index and pour point of lubricants [21–23]

		Mass fraction of different additives							
		0.02		0.03		0.04		0.05	
Sample	Base oil	VI	Pour point, (°C)	VI	Pour point, (°C)	VI	Pour point, (°C)	VI	Pour point, (°C)
S-1	SN70	102	−3	135	−7	189	−8	207	−12
	SN150	115	−6	142	−8	198	−9	220	−15
S-2	SN70	120	−6	147	−9	205	−12	221	−18
	SN150	125	−8	161	12	213	−15	232	−20
S-3	SN70	128	−6	169	−8	211	−9	229	−15
	SN150	132	−8	174	−10	223	−12	239	−15
S-4	SN70	136	−3	184	−7	218	−9	232	−9
	SN150	142	−6	201	−8	232	−9	255	−10
S-5	SN70	140	−3	199	−7	232	−9	240	−9
	SN150	146	−6	212	−6	234	−8	262	−10

3.3 Homo- and Copolymers of SBO and Sunflower Oil with MA and MMA as Multipurpose Additives (PPD, VII)

Derivatives of vegetable oils (like soybean oil (SBO), sunflower oil, etc.) have been used as additives as well as base stocks in the formulation of eco-friendly lubricant [14–18]. Li et al. [19] have shown the application of natural garlic oil as a high-performance and environmentally friendly extreme pressure additive in lubricants. Biresaw et al. [20] have mentioned the use of biobased polyesters as an extreme pressure additive in mineral oil. Erhan et al. in US Patent 7279448 B2 (2007) have described the use of poly(hydroxy thioether) vegetable oil derivatives as antiwear/antifriction additives for environmentally friendly industrial oils in automotive applications. The viscosity, thermal stability, and cold flow properties of epoxidized methyl oleate and epoxidized methyl linoleate, as well as that of the commercial epoxidized products, epoxidized soybean oil, and epoxidized 2-ethylhexyl soyate (Vikoflex™), have also been reported by Sharma et al. [24]. The application of nitrogen-containing derivatives of soybean oil as renewable feedstocks in diverse applications along with an alternative of petroleum-based lubricant has been reported [21]. Thus, there exists much research to develop environmentally benign additives based on vegetable oils, and due to ease of availability, relatively lower cost, and better performance, considerable attention has been given to research on soybean oil-based lubricants. Gobinda and Ghosh [25] have synthesized a homopolymer of SBO along with its copolymers with MA and MMA in two different concentrations (5% and 10%, w/w). The method with atom transfer radical polymerization (ATRP) uses anhydrous $FeCl_3$ as a catalyst, diethylenetriamine (DETA) as a ligand, azobisisobutyronitrile (AIBN) as an initiator, toluene as a solvent, and metallic "Fe" as reducing agent. The metallic "Fe" reduces Fe (III) to Fe (II) that controls the rate of polymerization. The prepared six different additives, viz., SBO,

Table 3.2 Biodegradability and pour point of the lubricants [21, 25]

	Biodegradability test, weight loss %						Pour point, °C				Pour point, °C			
	Days						SN70 additives, mass fraction				SN150 additives, mass fraction			
Additives	15	30	45	60	75	90	0.02	0.03	0.04	0.05	0.02	0.03	0.04	0.05
SBO	3	7	12	17	23	29	ND	ND	ND	ND	ND	ND	ND	ND
HSBO	2.5	5	9	14	22	28	−5	−6	−7	−8	−6	−8	−9	−11
CSBO/MA-5	2	2.5	6	11	15	18	−6	−9	−12	−18	−7	−9	−15	−18
CSBO/MA-10	1	2	5	8	14	17	−5	−7	−11	−14	−7	−9	−12	−14
CSBO/MMA-5	2	3	7.5	9	17	20	−5	−6	−9	−11	−6	−6	−9	−14
CSBO/MMA-10	2.5	3	7	10	16.5	19	−6	−6	−7	−11	−6	−8	−9	−12

homopolymer of soybean oil (HSBO), copolymer with 5% MA (CSBO/MA-5), copolymer with 10% MA (CSBO/MA-10), copolymer with 5% MMA (CSBO/MMA-5), and copolymer with 10% MMA (CSBO/MMA-10), have been characterized in terms of their performances in different mineral paraffinic base oils. Also, the biodegradability of each polymer has also been tested. The results of biodegradability and pour point of lubricants are given in Table 3.2. The main advantage of using vegetable-based additives over synthetic acrylates is the excellent biodegradability. The biodegradability test of the additives was performed by soil burial degradation test (SBT) method (ISO 846:1997) [26].

SBT Method A total of 1.5 g of each of the samples was taken to produce a film [19] and then buried in soil in a bacteriological incubator apparatus (Sigma Scientific Instruments Pvt. Ltd., Chennai, India) and thus subjected to the action of microorganisms in which soil is their major habitat. The soil sample was collected from North Bengal University campus having 25% moisture and pH 7.2. The experiments were conducted at 30 °C after each regular interval of 15 days up to 3 months at a relative humidity of 60%. The detail experimental procedure was given in the Karmakar and Ghosh [21].

Ultimate biodegradation is achieved when the substance is totally converted into carbon dioxide, water, mineral salts, and biotic mass (see Fig. 2.1). The extent of degradation was confirmed by measuring the percentage of weight loss (PWL) of the samples [27] and observing the shift in IR frequency of the ester carbonyl group after the test.

The addition of these newly developed biodegradable additives to mineral base stocks enhances the lubricant properties significantly. The control of polymerization through ATRP and the eco-friendly produces polymers having potential PPD, VII, and antiwear performances. Significant enhancement in VI and PP performances was observed by the incorporation of acrylates (MA and MMA) in the triglyceride backbone of soybean oil. Therefore, the synthesized additives would be a potential green alternative to the commercially available additives in the formulation of eco-friendly lubricant composition [25].

3.4 Homo- and Copolymers of Jojoba Oil as a Multipurpose Additive (PPD, VII)

Nassar et al. [29] prepared homo- and copolymers of jojoba oil and analyzed its significant lubricating properties, i.e., viscosity index improver and pour point depressant. The jojoba was polymerized as a homopolymer and novel six copolymers, and its features were compared with jojoba homopolymer. The results showed that the VI increases with increasing the chain length of alkyl units of both alpha olefins and acrylate groups; in other words, the pour point improved for additives based on alkyl acrylate. Liquid ester mixture of jojoba oil is refined from seeds of a desert shrub, Arizona, Northwestern Mexico, and Baja California [30]. The obtained jojoba oil differs from animal and vegetable oils in that it is not fat but has liquid wax. The extracted jojoba oil is distinctive among vegetable oils. Such type of vegetable oil has never before been available to industry in commercially usable quantities.

There are five essential characteristics that make them valuable for industrial applications:

(a) Its stability, molecular skeleton, and natural purity.
(b) It is having high resistance to oxidation.
(c) It can be reserved for several years without changing its chemical structure and properties.
(d) Its lubricating capacity.
(e) Its double bond nature [31].

Bisht et al. [32] studied the application of jojoba oil as an additive in vegetable oil base stocks as lubricants. The preliminary properties such as viscosity index (VI), rust protection, foaming, friction, and wear debris characteristics were estimated. Their results showed that blended jojoba oil could improve or impart certain desirable properties, such as VI improvement, antiwear, antirust, antifoam, and friction reduction properties [33]. Jojoba oil is identical with sperm whale oil which is wax liquid. It can be made up primarily of wax esters of docosenol (C22), eicosenol (C20), and eicosenoic acid (C20). It contains a little bit of saturated material, and the monounsaturated is the original constitution [34, 35]. The application of jojoba oil, with its longer chain length of 40–42 carbon atoms and high-polarity straight carbon chain, as a component of lubricant, reduces wear due to the strong adsorption of the polar molecules on the two metal surfaces (see Sect. 1.3 and Fig. 1.3). The viscosity of jojoba oils drops within the range of SAE 20, and the oil is able to be used as a base stock without dilution. The viscosity index is quite high compared with high-viscosity index (HVI) paraffinic base stocks which can improve lubrication especially at higher temperature [36]. Viscosity is a primary consideration in properly selecting lubricating oil for a given application. It should be high enough to provide good lubricating films but not so high that friction damages in the oil will be excessive. Meanwhile, vegetable oil viscosity varies with temperature; it is necessary to consider the actual operating temperature of the oil in the machine (see Table 7.5). Other concerns, such as whether equipment must be started at low temperatures, should be taken into account. The PPD acts via surface adsorption onto the wax crystals of the oil. The resulting surface layer of the PPD inhibits the growth of the wax crystals and their capacity to absorb the oil and form gel (see Figs. 4.1 and 4.6). In the absence of lengthy interconnecting crystals, oil can move freely throughout solid wax particles that are present [37]. By studying the literature survey on using jojoba as an additive in lubricants, one can find that jojoba oil was taken as it is with different % as viscosity index improvers [34, 35] or by blending a pour point depressant to develop flow properties of jojoba imparted with lubricating oil.

3.5 Natural Garlic Oil as an Extreme Pressure Additive

Natural garlic oil (NGO) is extracted from biodegradable vegetable esters. It is a blend of a garlic-rich organosulfur compound (OSC) complex, which is withdrawn from cloves of garlic. Generally, plenty of garlic OCSs are obtained from alliin via the action of alliinase and followed by the rearrangement reaction. Finally, garlic oil is extracted by steam distillation to condense the volatile OSCs in garlic [38]. The major OSCs in NGO appear as diallyl disulfides that are analogous to sulfurized isobutylene (SIB) chemical skeleton [39]. Therefore, this is the reason to believe that NGO may offer good lubricant load-carrying capacities. In lubricating oil, NGO as an extreme pressure additive had so far not been studied abundantly, very few literature is available for natural garlic oil as lubricant additives, for example, Li et al. [19] reported the NGO as high-performance (environmentally friendly), extreme pressure additive in lubricants. The extreme pressure property of different PAO10 and synthetic hydrocarbon base fluids with NGO and SIB as additive is evaluated by four-ball tester. The four-ball test results are shown in Table 3.3 and have shown that 1 wt. % NGO incorporation into the base stocks may possibly develop the weld point of the base fluids from 126 to 800 kgf or more. Moreover, the extreme pressure results established that NGO can provide superior load-carrying capacity in the selected base stocks than the conventional extreme pressure additive, i.e., SIB. Also, X-ray photoelectron spectroscopy (XPS) results showed that NGO and SIB experienced a similar tribochemical process with the generation of tribofilms composed of iron oxides, iron sulfates, iron sulfide, etc. The experimental results and literature showed NGO can be used as an effective, eco-friendly, extreme pressure additive in environmentally sensitive areas. The researchers [19, 38, 39] have reported GC-MS analysis results, which reveal that the composition of natural garlic oil is complex. Wear scar results showed that the weld points of the fluids were significantly improved with the addition of NGO and SIB. Concurrently

Table 3.3 Extreme pressure property of different wt. % of SIB and NGO additive on base oil [19, 38, 39]

	EP (weld point), kgf									
	SIB additive concentration (wt %)					NGO additive concentration (wt %)				
Base oil	0	0.25	0.5	0.75	1.0	0	0.25	0.5	0.75	1.0
PAO10	126	250	250	315	315	126	250	315	500	800
PAG660	160	–	–	–	315	–	–	–	–	800
3970	126	–	–	–	315	–	–	–	–	800
Rapeseed oil	126	250	315	315	315	126	250	400	800	>800

the author studied the NGO performance with different base fluids PAO, synthetic esters, vegetable oils, and polyalkylene glycols; the results showed that NGO could enhance the load-carrying capacity (see Table 3.3). Additionally, the four-ball test results proved that NGO exhibits superior extreme pressure properties than SIB under the same experimental conditions. XPS results verified that a tribochemical process involves the formation of tribofilms (iron sulfide) formed between the friction pairs. By these results, it is to be believed that NGO is a promising alternative to SIB to be used as an eco-friendly EP additive.

3.6 Palm Oil Methyl Ester as an Antiwear Additive

Maleque et al. [40] published a review article on a case study of palm oil methyl ester (POME) as an additive, and they reported the specific lubricating properties of a vegetable-based lubricant additive and its application. However, environmentally friendly natural esters of vegetable oils can propose advantages on natural resource renewability and biodegradability and also concurrently display acceptable performance in a variety of applications. Initial research stated that the best boundary additives have surfactant, i.e., long-chain organic molecules with an active polar end group [41]. For that reason, it can be said that POME composition can deliver adequate boundary lubrication due to the existence of a polar group and can act as an antiwear additive for the lubrication process. Masjuki and Maleque [42, 43] have reported four-ball wear test results using 0, 3, 5, 7, and 10 volume percent of POME with mineral oil. The results showed 5% POME could perform as an antiwear additive. The palm oil methyl ester is extracted from crude palm oil through transesterification reaction and has sulfur content (0.002 wt%); hence, it behaves as eco-friendly lubricant additives. It can be assumed that POME composition can provide useful boundary lubrication, due to the polar structure dissipation on nonpolar base lubricant, and it may act as an antiwear additive and friction modifier (see Fig. 4.4). Liew et al. [44] found that mineral oil with POME (additive) is having shorter running-in time and lower steady-state frictional coefficient at slight loads of around 600 and 800 N. Finally, the authors concluded that the difference in the friction coefficient generated in mineral oil with and without POME became more detectable at pressures above 800 N.

3.7 Lipoate Esters as a Multipurpose Additive (VII, AO, EP)

Biresaw et al. [45, 46] synthesized seven lipoate esters by esterification of lipoic acid with different structures of the hydroxyl group in the presence of a solid acid catalyst and without solvent. Four of them displayed good solubility in high oleic sunflower oil (HOSuO) and polyalphaolefin (PAO-6) base oils. Additive properties were investigated by varying the additive concentrations in the range 0–20% (w/w). The author focused on the additive properties which are density, kinematic viscosity, viscosity index (VI), oxidation stability from onset and peak oxidation temperatures on a pressure differential scanning calorimetry (PDSC), antiwear (AW), and extreme pressure (EP) from four-ball tester. Neat lipoate esters, i.e., homopolymer, displayed excellent oxidation stability and high VI. As additives in HOSuO base oil, lipoate esters exhibited good viscosity index improver (VII), good antioxidant (AO), and good extreme pressure (EP), but poor AW properties. The VI and VII properties were identified by the presence of low concentrations of polymeric by-products. It determines that lipoate esters can be developed into useful multifunctional and multipurpose biobased lubricant additives with excellent VII, AO, and EP properties. The alcohols differed in their molecular weight and complexity of chain branching (none or linear, straightforward, and complex). Four of the seven lipoate esters, viz., 2-ethylhexyl, n-octyl, n-dodecyl, and isostearyl, which displayed good solubility in HOSuO and PAO6, were additionally investigated for their lubricant additive properties. Table 3.4 shows the lubrication properties of HOSuO with different % of lipoate ester as additives.

Neat lipoate esters as base oil:

1. Displayed density close to 1 g/ml at 40 °C, which decreased with increasing chain length and increasing temperature.
2. Had kinematic viscosity in the range 74–53,714 cSt at 40 °C; kinematic viscosity increased with increasing concentration of polymeric by-product in the order isostearyl < n-octyl <2-ethylhexyl < n-dodecyl.
3. Displayed high VI in the range 70–567; VI increased with increasing concentration of polymeric by-product in the order isostearyl < n-octyl <2-ethylhexyl < n-dodecyl.

Table 3.4 Oxidation stability[a] and EP weld point of HOSuO with lipoate ester additives [45, 46]

	Onsite temperature (°C) of blends with lipoate ester (%w/w)					Four-ball EP weld point (kgf) of lipoate ester in HOSuO (% w/w)				
Lipoate ester	0[b]	1	5	10	20	0[c]	1	5	10	20
2-Ethylhexyl	187	192	200	212	221	120	220	320	320	320
n-Octyl	187	192	200	211	221	120	240	320	420	440
n-Dodecyl	187	194	202	212	220	120	200	260	440	480
Isostearyl	187	190	198	204	218	120	220	300	380	420
TPS-32	187	185	199	200	201	120	–	–	–	–

[a]ASTM D6186, [b,c]neat HOSuO data from [46], TPS-32: di-tert-dodecyl polysulfide

4. Displayed oxidation stability that is independent of chain structures and much higher (onset temperature = 226–229 °C; peak temperature = 247–253 °C) than that for neat HOSuO (onset temperature (OT) = 187 °C; peak temperature (PT) = 200 °C).

   Lipoate esters as additives:

1. Displayed solubility properties in polar (HOSuO) and nonpolar (PAO6) base oils that are consistent with expectations based on their chain structure (chain length and degree of branching).
2. Increased the density of HOSuO at 40 °C.
3. Increased the kinematic viscosity of HOSuO at 40 °C; the relative increase in kinematic viscosity is a function of the concentration of polymeric by-product in the ester.
4. Displayed good viscosity index improver (VII) properties, which resulted in increased VI of HOSuO with increased concentration; VII properties were attributed to the presence of polymeric by-products in the product mixture.
5. Displayed excellent antioxidant (AO) properties which, except isostearyl lipoate, resulted in increased oxidation stability of HOSuO with increasing lipoate ester. AO properties were independent of lipoate ester structure, as demonstrated by increases in OT and PT with increasing [lipoate ester]. It varied in the range 0–20% w/w which displayed poor antiwear properties where, at 1% [lipoate ester], a significant increase of both coefficients of friction (COF) and wear scar diameter (WSD) of HOSuO was observed. The effect of 1% additive on COF is independent of lipoate ester structure; but the structure had an effect on WSD that, at 1% lipoate ester, increased in the order: 2-ethylhexyl < n-octyl < n-dodecyl < isostearyl.
6. Displayed excellent EP properties that increased the weld point of HOSuO blends from 120 kgf without additive to 320–480 kgf at [lipoate ester] of 20% (w/w). EP properties are independent of the ester chain structure but a result of the S–S structure of the lipoate ester. This research demonstrates that lipoate esters can be developed as multifunctional and multipurpose biobased lubricant additives with excellent VII, AO, and EP properties.

3.8 Rapeseed Oil as a Friction Modifier Additive

In recent days, vegetable oils are possible alternative to mineral oil as base stocks in lubricants (see Fig. 1.5). They typically have an easy biodegradable nature and low toxicity, and their resources can be recycled [47]. Vegetable oils show excellent tribological properties such as friction reduction, wear control, improved extreme pressure, etc., when used as base stocks or additives. Durak [48] studied the performance of lubricants which were prepared mineral oil (SAE 20 W50) with different amounts of Turkish-originated rapeseed oil (RSO). The author tried to conclude the result on the friction coefficient under static loading. RSO is an environmentally friendly, renewable, and biodegradable resource (see Tables 2.5 and 2.6). The application of RSO as a lubricants and additive is an important alternative to mineral oil. Due to it being edible, RSO is less favored than other vegetable oils. Durak [48] studied the RSO addition effects of oil on the friction coefficient in the journal bearing at 25 and 100 °C. The mineral oil with RSO additive acts in decreasing the friction coefficient at high speeds and even at medium loads. From the results, RSO performed greater reduction in the coefficient of friction at 25 °C than at 100 °C. At ≥50% concentration ratio, maximum reduction rates were obtained with the usage of lubricant with RSO. The theoretical and experimental study revealed RSO could be used as friction modifier at broad range of temperature conditions.

3.9 Conclusion

Lubrication performances of homo- and copolymers as additives which are derived from different types of natural esters of vegetable oil are reviewed. The homo- and copolymers have specific properties to enhance the PPD, VII, AO, EP, and AW when blended with suitable base oils. The copolymers of SBO with 1-decene and styrene perform as better viscosity index improvers. Copolymers of SBO with methyl acrylate and 1-decene function as better antiwear additives. The soy-methyl acrylate copolymer has a better pour point depressant property. The other case reported that significant enhancement of VI and PP performances was observed by the incorporation of acrylate polymerization through ATRP (MA and MMA) in the triglyceride backbone of soybean oil. The homo- and copolymers of jojoba were assessed as viscosity index improvers and pour point depressants for lubricating oil. NGO was used as an environmentally friendly extreme pressure additive in lubricating oils. Palm oil methyl ester showed better performance as an extreme pressure additive. The lipoate esters displayed as right viscosity index improver (VII), good antioxidant (AO), and good extreme pressure (EP) but poor AW properties. Moreover, due to being biodegradable, all the additives discussed in this chapter are eco-friendly.

References

1. Pedersen KS, Rønningsen HP (2003) Influence of wax inhibitors on wax appearance temperature, pour point, and viscosity of waxy crude oils. Energy Fuel 17:321-–328
2. Florea M, Catrinoiu D, Luca P, Balliu S (1999) The influence of chemical composition on the pour-point depressant properties of methacrylate copolymers used as additives for lubricating oils. Lubr Sci 12:31–44
3. Onyeji LI, Aboje AA (2011) The effect of additive on the viscosity index of lubricating oil (engine oil). Int J Eng Sci Technol 3:1864–1869
4. Azim AAAA, Nassar AM, Ahmed NS, Kamal RS (2006) Preparation and evaluation of acrylate polymers as pour point depressants for lube oil. Pet Sci Technol 24:887–894

5. Mohamed MM, El-Naga HHA, El Meneir MF (1994) Multifunctional viscosity index improvers. J Chem Technol Biotechnol 60:283–289
6. Liu W, Hu L, Zhang Z (1995) Friction and wear of the film formed in the immersion test of oil containing antiwear and extreme pressure additives. Thin Solid Films 271:88–91
7. Alemán-Vázquez LO, Villagómez-Ibarra JR (2001) Polyisobutenylsuccinimides as detergents and dispersants in fuel: infrared spectroscopy application. Fuel 80:965–968
8. Nassar AM, Ahmed NS, Kamal RS, Abdel-Azim AA, El-Nagdy EI (2005) Preparation and evaluation of acrylate polymers as viscosity index improvers for lube oil. Pet Sci Technol 23:537–546
9. Ghosh P, Das M, Upadhyay M, Das T, Mandal A (2011) Synthesis and evaluation of acrylate polymers in lubricating oil. J Chem Eng Data 56:3752–3758
10. Erhan SZ, Asadauskas S (2000) Lubricant base stocks from vegetable oils. Ind Crop Prod 11:277–282
11. Adhvaryu A, Erhan SZ (2002) Epoxidized soybean oil as a potential source of high-temperature lubricants. Ind Crop Prod 15:247–254
12. Mercurio P, Burns KA, Negri A (2004) Testing the ecotoxicology of vegetable versus mineral based lubricating oils: 1. Degradation rates using tropical marine microbes. Environ Pollut 129(2):165–173
13. Mofijur M, Masjuki HH, Kalam MA, Shahabuddin M, Hazrat MA, Liaquat AM (2012) Palm oil methyl Ester and its emulsions effect on lubricant performance and engine components wear. Energy Procedia 14:1748–1753
14. Sharma BK, Doll KM, Erhan SZ (2007) Oxidation, friction reducing, and low-temperature properties of epoxy fatty acid methyl esters. Green Chem 9(5):469–474
15. Durak E, Karaosmanoglu F (2004) Using cotton seed oil as environmentally accepted lubricant additive. Energy Sources 26:611–625
16. Boshui C, Nan Z, Kai L, Jianhua F (2012) Enhanced biodegradability, lubricity and corrosiveness of lubricating oil by oleic acid Diethanolamide phosphate. Tribol Ind 34(3):152–157
17. Biswas A, Adhvaryu A, Stevenson DG, Sharma BK, Willet JL, Erhan SZ (2007) Microwave irradiation effects on the structure, viscosity, thermal properties and lubricity of soybean oil. Ind Crop Prod 25(1):1–7
18. Liu Z, Sharma BK, Erhan SZ, Biswas A, Wang R, Schuman TP (2015) Oxidation and low temperature stability of polymerized soybean oil-based lubricants. Thermochim Acta 601:9–16
19. Li W, Jiang C, Chao M, Wang X (2014) Natural garlic oil as a high-performance, environmentally friendly, extreme pressure additive in lubricating oils. ACS Sustain Chem Eng 2:798–803
20. Biresaw G, Asadauskas SJ, McClure TG (2012) Polysulfide and Biobased extreme pressure additive performance in vegetable vs Paraffinic Base oils. Ind Eng Chem Res 51:262–273
21. Karmakar G, Ghosh P (2015) Soybean oil as a biocompatible multifunctional additive for lubricating oil. ACS Sustain Chem Eng 3:19–25
22. Kuzmić AE, Radošević M, Bogdanić G, Sricá V, Vuković R (2008) Studies on the influence of long chain acrylic esters polymers with polar monomers as crude oil flow improver additives. Fuel 87:2943–2950
23. Soldi RA, Oliveira ARS, Barbosa RV, César-Oliveira MAF (2007) Polymethacrylates: pour point depressants in diesel oil. Eur Polym J 43(8):3671–3678
24. Biswas A, Sharma BK, Willett JL, Erhan SZ, Cheng HN (2008) Soybean oil as a renewable feedstock for nitrogen-containing derivatives. Energy Environ Sci 1:639–644
25. Karmakar G, Ghosh P (2016) Atom transfer radical polymerization of soybean oil and its evaluation as a biodegradable multifunctional additive in the formulation of eco-friendly lubricant. ACS Sustain Chem Eng 4:775–781
26. Chandure AS, Umare SS (2007) Synthesis, characterization and biodegradation study of low molecular weight polyesters. Int J Polym Mater 56:339–353
27. Ghosh P, Karmakar G (2014) Evaluation of sunflower oil as a multifunctional lubricating oil additive. Int J Ind Chem 5(7)
28. Liu M, Huang Z, Yang Y (2010) Analysis of biodegradability of three biodegradable mulching films. J Polym Environ 18:148–154

29. Nassar AM, Ahmed NS, Nasser RM (2015) Jojoba polymers as lubricating oil additives. Pet Coal 57(2):20–129
30. Miwa TK (1984) Structural determination and uses of jojoba oil. JAOCS 61(2):407–410
31. James HB, Harry O (1982) Isomerization of jojoba oil and products thereof, US patent 4.329.298
32. Bisht RPS, Sivasan Karan GA, Bhatia VK (1993) Additive properties of jojoba oil for lubricating formulations. Wear 161:193–197
33. Anand ON, Chhibber VK (2006) Vegetable oil derivatives: environment–friendly, lubricants and fuels. J Synth Lubr 23:91–107
34. Allawzi M, Abu-Arabi MK, Al-Zoubi HS, Tamimi A (1998) Physicochemical characteristics and thermal stability of Jordanian jojoba oil. JAOCS 75(1):57–62
35. Gisser H, Messina J, Chasan D (1975) Jojoba oil as a sperm oil substitute. Wear 34:53–63
36. Sivasankaran GA, Bisht RPS, Gupta VKM, Sethuramiah A, Bhatia VK (1998) Jojoba–oil–based two stroke gasoline engine lubricant. Tribol Int 21(6):327–333
37. Rizvi SQA (2009) A comprehensive review of lubricant chemistry, technology, selection and design. ASTM International, West Conshohocken, pp 100–211
38. Chin-Chung W, Lee-Yan S, Haw-Wen C (2001) Differential effects of garlic oil and its three major organosulfur components on the hepatic detoxification system in rats. J Agric Food Chem 50(2):378–383
39. Weinberg DS, Manier ML, Richardson MD, Haibach FG (1993) Identification and quantification of organosulfur compliance markers in a garlic extract. J Agric Food Chem 41(1):37–41
40. Maleque MA, Masjuki HH, Sapuan SM (2003) Vegetable-based biodegradable lubricating oil additives. Ind Lubr Tribol 55(3):137–143
41. Williams JA (1994) Engineering tribology. Oxford University Press, Oxford. 166–8: 348–58
42. Masjuki HH, Maleque MA (1997) Wear, performance and emissions of a two-stroke engine running on palm oil methyl ester blended lubricant. Proc. Imech E, Part J J Eng Tribol 210:213–219
43. Masjuki HH, Maleque MA (1997) Investigation of the anti-wear characteristics of palm oil methyl ester using a four-ball tribometer test. Wear 206:179–186
44. Liew WYH, Dayou S, Dayou J, Siambun NJ, Ismail MAB (2014) The effectiveness of palm oil methyl ester as lubricant additive in milling and four-ball tests. Int J Surf Sci Eng 8:153–172
45. Biresaw G, Compton D, Evans K, Bantchev GB (2016) Lipoate Ester Multifunctional Lubricant Additives. Ind Eng Chem Res 55:373–383
46. Biresaw G, Laszlo JA, Evans KO, Compton DL, Bantchev GB (2014) Tribological investigation of lipoyl glycerides. J Agric Food Chem 62(10):2233–2243
47. Cao Y, Yu L, Liu W (2000) Study of the tribological behavior of sulfurized fatty acids as additives in rapeseed oil. Wear 244:126–131
48. Durak E (2004) A study on friction behavior of rapeseed oil as an environmentally friendly additive in lubricating oil. Ind Lubr Tribol 56(1):23–37

Chapter 4
Biodegradable Polymers as Lubricant Additives

Abstract The physical, chemical, and antiwear performance of lubricant which contains a biodegradable polymer as additive is literature reviewed. Readily available and with its broad range of applications, vegetable oils are reviewed as a very important class of renewable resources for lubricants as well as its additives. The readily available and modified biodegradable polymers are used as renewable novel additives in the lubricant and is reviewed and discussed here.

Keywords Biodegradable polymers • Antiwear additive • Extreme pressure additive • Biobased polyesters • Copolymers • Homopolymer

4.1 Introduction

Nowadays, due to developing environmental responsibility, vegetable oils are finding a new application in lubricants in both transportation and industrial applications. They can furnish the important environmental benefits on fuel consumption and biodegradability [1]. Nevertheless, both the thermal and oxidative instabilities are the major limitations of vegetable-based natural ester fluids (see Sect. 1.3 and Fig. 1.6). Keeping this as basic literature references and the current studies toward the development of lubricating additives and base oil, as well as lots of challenges has been made with polymers additive to develop the physical, chemical, and other characteristics. Polymeric additives have been used as pour point depressants to reduce the pour point and increase the viscosity index of lubricants [2–6]. Pour point (PP) is the lowest temperature at which lubricants can flow and below which there is no movement in it. Since the performance of such additives in field conditions is very much dependent on the chemical skeleton and structural morphology of such polymers in the preferred medium [7], viscometric studies in dilute solutions give the information about the quality of the base stock employed and molecular chain conformation in dilute solution. In the meantime, research publications regarding such information are very few [8, 9]; the recent and present research comprises a viscometric study of the homo- and copolymers. The polymer additives of numerous types are performed in lubricant composition to improve the viscosity,

D. Gnanasekaran, V.P. Chavidi, *Vegetable Oil based Bio-lubricants and Transformer Fluids*, Materials Forming, Machining and Tribology,
DOI 10.1007/978-981-10-4870-8_4

pour point, antiwear, oxidation stability, and other properties of the lubricant (see Tables 3.1, 3.2, 3.3, and 3.4). Additionally, they have been providing performance features such as improved low-temperature flow behavior, dispersion, and stiffening. Even though the modern claim for eco-friendly technology directed us to chemically blend the vegetable oil with polymers through the process of copolymerization in anticipation of producing an ideal blend of performance as well as eco-friendly lubricants, hence this chapter covers the additive as biodegradable polymers in vegetable base oil as lubricants. The varieties of vegetable oil-based polymers have been prepared by a free radical, cationic, olefin metathesis and condensation polymerization reaction mechanism. The obtained materials have shown a broad range of thermophysical and mechanical properties from soft and elastic rubbers to hard and stiff plastics, which show potential as substitutions for petroleum-based plastics. The effects of biodegradable polymers as an additive in various aspects of lubricants are discussed as follows.

4.2 Oleates, DIDA, PAO-2, and PMMA as Pour Point (PP) Depressant

The dilution effects of biodegradable oils, i.e., polyalphaolefin (PAO-2), diisodecyl adipate (DIDA) and oleates with base oil and the impact as a pour point depressant (PPD) were investigated and reported by Asadauskas and Erhan [10]. Meanwhile, the freezing or crystallization of mixed saturated and unsaturated triacylglycerols was a complex thermodynamic mechanism; the study was narrow to detect pour point (PP). Increasing saturation and molecular weight of vegetable oils displayed higher pour point, but at the same time, *Cis* unsaturation and hydroxy groups preferred to lower the pour point value. The author [10] finds out the dilution effect and report seemed to be less active than dilution with DIDA and PAO-2. The addition of 1% (w/w) PPD reduced the PP to −33 °C for canola and −24 °C for high oleic sunflower oil (HOSO). Though, the pour point further depressed by neither higher the amounts of PPD nor the incorporation of diluent, the pour point depression has not directly related to the concentration of diluent and ceased with further dilution. The low-temperature behavior of natural esters restricts their prospect as biodegradable lubricants; however, the well-adjusted usage of PPD with diluents is to be permitted to convey some developments. They reported the changes in PP, while vegetable oils were blended with diluents. Castor oil demonstrates that the PP, particularly lower than those of soybean oil (SOY), HOSO, and canola oils (CO), advises that the interaction by H-bonding between hydroxyl groups of ricinoleic acid interferes. The author have tried an attempt to observe effects on PP, oleates, DIDA, and PAO-2 (20%) mixed with HOSO and soybean oils. The pour point of those various types of biodegradable lubricants is shown in Table 4.1. From the Table 4.1, we can understand the oleates depressed the PP slightly; though, DIDA and PAO-2 showed greater effects. Considerably, a larger portion of the diluent does not provide balanced depressions of PP. Even though the difference is not

Table 4.1 Pour point of different types of vegetable oil with different diluents [10]

	SOY, pour point, °C					HOSO, pour point, °C					Canola oil, pour point, °C				
	W/W % of diluent					W/W % of diluent					W/W % of diluent				
Diluent	20	0	0.4	1	2	20	0	0.4	1	2	20	0	0.4	1	2
2-Ethoxy oleates	−12					−18									
Isobutyl oleates	−12					−15									
TPT	−12					−18									
PT	−12					−15									
DIDA	−18					−21									
PAO	−21					−21									
PMMA		−9	−18	−18	−18		−12	−21	−24	−24		−18	−30	−33	−33

TPT Trimethylolpropane trioleate, *PT* Pentaerythritol tetraoleate, *PMMA* Poly(methyl methacrylate)

Table 4.2 Effect of PAO-2 and PPD on pour point of HOSO [11]

Blend ratios (w/w)			
HOSO	PAO-2	PPD % (w/w)	Pour point °C
0	100	0	−72
100	0	0	−12
90	10	0	−15
80	20	0	−21
60	40	0	−24
90	10	0.66	−24
80	20	0.66	−24
60	40	0.66	−27
100	0	1	−24
90	10	1	−24
80	20	1	−27
60	40	1	−27

statistically important in the dilution range of 0–80%, DIDA, being more branched than PAO-2, seems to slow down the solidification more efficiently.

The PAO-2 does not generate an effect as would be estimated from its PP of −72 °C. Mainly, during the establishing of the pour point, solidification of the entire bulk liquid was noticed than the sedimentation of hard particulates. In that condition, particulates were not precipitating even at 80% dilution. At high dilutions, 90% of PAO-2, the solidification mechanism could change resulting in a drop of PP. Various amounts of PPD appear to exert a more pronounced effect as suggested by the PP of canola oil, HOSO, and SOY. The 0.4 wt.% of PPD considerably diminishes the degrees of solidification but is somewhat more active on canola oil. Increasing the concentration of PPD may still slow down the crystallization; however, the further depression cases are somewhat fast. Compared to diluents, PPD gives a quite stronger effect on reducing the degrees of crystallization. Since the two mechanisms of PPD may be different, the influence of the combination of both diluent and PPD was investigated, and results are shown in Table 4.2. However, there is no synergism between the two mechanisms of depressing PP. The results seem that even 40% of PAO-2 does not depress the PP of HOSO more significantly than does 1% of PPD, and the blends still demonstrate PP nearer to HOSO than those of the diluents. Thus, both the diluents and PPD provided a little bit of development in the low-temperature behavior.

4.3 Homo- and Copolymers of Sunflower Oil as VII, PPD, and AW Additive

The chemical background of biodegradable multifunctional additive was derived from vegetable oil (sunflower oil) by Ghosh and Karmakar [11]. Two different ways performed the solvent-free synthesis of a homopolymer of sunflower oil (SFO);

microwave irradiation (MI) method and thermal technique used benzoyl peroxide as an initiator. The author performed the evaluation of pour point depressant (PPD), viscosity index improver (VII), and antiwear (AW) in the different nature of base oils [BO1 and BO2 (supplied by IOCL)] with SFO as an additive carried out by ASTM methods. Table 4.3 explained the comparison of their performances. The microwave-assisted (MI) method provides better VI and PP values as compared to the thermally synthesized additive. Thus, polymerized SFO could be used as a suitable biodegradable lube oil additive, and therefore the MI method may be considered as a cost-economical greener approach to synthesis of lube oil additive.

The properties of PPD, AW, and VI of lubricant contain an additive which is synthesized by the MI method and is better than the thermal method. Thermal stability of P-2 (a homopolymer of SFO prepared by MI method) is greater as compared to P-1 (thermally prepared SFO polymer). Even though both of them are identified as more or less equally effective as thickeners for the base oils studied, in general, it is detected that viscosity index, thickening power, AW property, and PP increase with increasing concentration. This research also explains that VI values of the additive-doped base oils can be influenced the base oil composition. Both of the additives displayed efficient biodegradability by a soil burial test and disk diffusion tests. They were not projected for commercial acrylate-based additives, and this is the main advantage of SFO additives. Therefore, it may be determined that the most economical MI method for synthesis of SFO is not only a greener cost-effective approach but also one that is better performing.

Gosh et al. [12] synthesized and characterized lubricants which have copolymers of sunflower oil and methyl methacrylate and decyl acrylate as an additive for viscosity index improver. The prepared additives were doped with base oils (sunflower and mineral oils), and then the VI of both were compared. The author concluded the VI of the lubricants depends on the nature of base oils and the type and concentration of VII. Gosh et al. [12] have prepared, for a comparison, analyzed, and evaluated in the comparable fashion the respective homopolymer, e.g., polymer of sunflower oil, poly(methyl methacrylate), and poly(decyl acrylate). Overall, VI increases with the increase in the additive concentration in solution irrespective of the nature of the virgin oils. Again the VI values of the copolymers doped with virgin oils are slightly higher in comparison to the homopolymer doped with base oils, and the prediction is not dependent on the nature of the base oils.

4.4 Ethylene-Vinyl Acetate and Ethyl Cellulose as VII and PPD Additive

Low-temperature performance is one of the primary limitations regarding the use of vegetable oils as biolubricants more than mineral or synthetic oil lubricants. Quinchia et al. [13] studied the low-temperature behavior of different types of vegetable oil base stocks of greasing uses; also, their composites with various viscosity

Table 4.3 Viscosity index, pour point, antiwear, and thickening performance evaluation of lubricants [11, 12]

		% (W/W) of the additives and its properties														
		2%				3%				4%				5%		
Samples	Base oils	THK	PP (°C)	VI	WSD (mm)	THK	PP (°C)	VI	WSD (mm)	THK	PP (°C)	VI	WSD (mm)	THK	PP (°C)	VI
P-1	BO1	2.7	−6	113	0.940	3.8	−9	126	0.937	5.17	−12	131	0.961	6.2	−15	131
	BO2	1.01	−9	101	0.982	1.56	−12	127	0.978	2.19	−15	163	0.973	3.13	−15	205
P-2	BO1	2.84	−6	121	0.938	3.9	−9	131	0.927	5.4	−15	128	0.921	6.8	−18	131
	BO2	1.09	−9	103	0.976	1.6	−12	130	0.968	2.25	−15	164	0.964	3.19	−18	207

WSD Wear scar diameter (mm), Load 147 ± 2 N, *THK* Thickening values

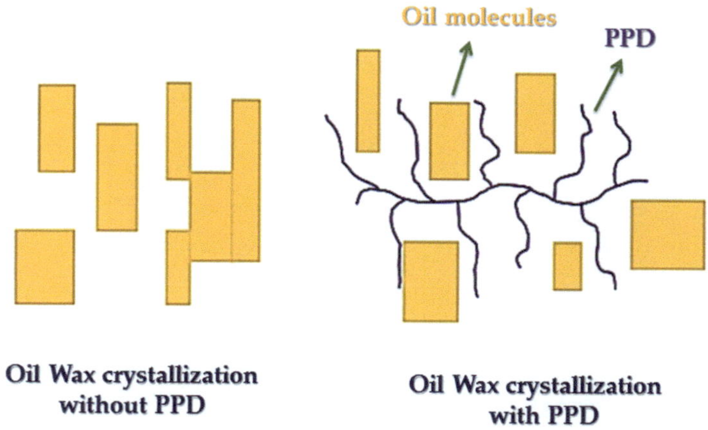

Fig. 4.1 Type of PPD interactions with vegetable oil (base oil) molecules

improvers and pour point additives were investigated by pour point calculation, thermal analysis (DSC), and viscosity measurements at different low-temperature conditions. The low-temperature properties of vegetable oil-based lubricants are predominantly influenced by the amount of polyunsaturated fatty acids (PUFAs). The PPD additives are used to express the positive effect by lowering the pour point and increasing the low-temperature character of the natural esters of vegetable oils. The general interaction of PPD additive with the vegetable oil (base oil) molecule is shown in Fig. 4.1. The obtained results were found to be dependent on the fatty acid structure of the vegetable oil. In this scenario, the outstanding results were observed by the sunflower oil (SFO)/PPD composite which reached a pour point of −36 °C in comparison to −18 °C for the neat oil. By contradistinction, the worst result was obtained from the HOSO/PPD composite (−21 °C) in comparison with HOSO (−18 °C). On the other hand, it was established that the ethyl cellulose (EC), used as a viscosity modifier, encourages a delay in HOSO crystallization, generating a comparable result than PPD tested, besides increasing the viscosity. The effects of two different types of environmentally friendly viscosity improvers, i.e., EC and EVA [14, 15], on the low-temperature flow behavior of HOSO were verified.

Although EVA is used effectively as a wax crystal convertor in mineral oils [16], the obtained results indicated that EC had a positive effect in reducing HOSO crystallization temperature (similar to PPD), but EVA has an entirely contrasting effect. Additionally, as can be seen in Table 4.4, EVA brings an extra inflection point, around 13 °C, with an undesirable enhancement in HOSO viscosity. In addition to this, low-temperature flow behavior of ternary blends of HOSO, a viscosity modifier, and PPD is also discussed. The results in Table 4.4 showed the blend containing EC (HOSO/EC/PPD) yields similar pour point (−21 °C) and no synergy was found. However, EVA blend, i.e., HOSO/EVA/PPD, showed intermediate pour point results in between those observed for HOSO/EVA and HOSO/PPD blends. Hence, it is better to use EC as a potential additive in the formulation of eco-lubricants having HOSO, since it improves lubricant viscosity, viscosity index, and pour point

Table 4.4 Different types of vegetable base oil and biodegradable polymer additives on viscosity, VI, and pour point [13–20]

	Lubricants																		
	PPD 0%					PPD 1%					EVA 1%	EVA 2%	EVA 3%	EVA 4%	EC1% + PPD1%	EVA 4% + PPD1%	SBS 2%	SBS 3%	SBS 4%
Properties	HOSO	SO	CO	SYO	RO	HOSO	SO	CO	SYO	RO	HOSO	HOSO	HOSO	HOSO	HOSO	HOSO	HOSO	HOSO	HOSO
K.V. 40 °C, cSt	38	37	242	35	36	40	34	196	37	38	63	87	128	168	49	177	73	113	145
K.V. 100 °C, cSt	10	9	21	8	8	10	9	20	9	10	13	16	22	29	15	32	15	23	27
VI	257	248	116	250	240	250	263	118	237	265	Nil	Nil	Nil	218	332	226	Nil	Nil	Nil
Pour point, °C	−18	−18	−21	−18	−21	−21	−36	−33	−33	−33	Nil	Nil	Nil	−12	−21	−18	Nil	Nil	Nil

depressant behavior significantly. The vegetable oils with lesser unsaturated/saturated fatty acids ratio crystallize at higher temperatures and, on the contrary, that the concentration of PUFAs has, even more, influence on low-temperature performance than the concentration of saturated fatty acids. Therefore, the rapeseed oil (RO) performs better at low temperatures than with a similar molecular backbone structure, i.e., HOSO, soybean oil (SOY), and sunflower oil (SO). With the results obtained in their study, the author made the conclusion that the PPD improves the low-temperature behavior of all vegetable oils studied. EC as an additive is used as viscosity modifiers and also encourages a delay in HOSO crystallization, further increasing the viscosity.

Quinchia et al. [17] studied the frictional and lubricating thin film-forming properties of several types of improved natural esters of vegetable oil-based lubricants [high oleic sunflower oil (HOSO), soybean oil (SOY), and castor oil (CO)], using 4 wt% of EVA copolymer and 1 wt% of ethyl cellulose as additives. The viscosity, viscosity index, and pour point of the different formulated biodegradable lubricants are shown in Table 4.4. It would be assigned to its hydroxyl group that increased both the viscosity and polarity of natural ester fluids. EVA copolymer utilized a slight effect on the film-forming character, mainly reducing the friction and wear in the blended lubricated area. On the other hand, the author concluded the ethyl cellulose is considerably extra efficient with castor oil, in improving both mixed and boundary lubrication.

4.5 Ethylene-Vinyl Acetate and Styrene-Butadiene-Styrene (SBS) as a VII Additive

Quinchia et al. [14] reported HOSO was blended with additives at a concentration range between 1 and 5% (w/w). Blends were mixed by stirring, at 300 RPM and agitation times for 5–10 h, at 100–150 °C, depending on the polymer and its nature. Thermal treatment was necessary to solubilize the polymer completely in HOSO. A homogeneous phase was obtained in all cases, although a certain degree of aggregation was detected with aging for polymer ratio > 3% (w/w) of the SBS/HOSO blends. The reported work [14] was to develop the novel formulations of environmentally friendly lubricant with enhanced viscosity thermal susceptibility and kinematic viscosity. With this aim, the polymer of HOSO was mixed with polymeric additives, such as EVA and SBS copolymers, at various concentrations 1 to 5% (w/w). The chemical structure of both vinyl acetate and styrene-butadiene-styrene is shown in Fig. 4.2. Both EVA and SBS copolymers could be reasonably used as additives to increase the viscosity of high oleic sunflower oil, thus improving the low viscosity values of the oil. HOSO viscosity increases with polymer concentration. Unambiguously, the blends of EVA/HOSO exhibit higher viscosity values, which are desirable for applications of bearings and four-stroke engine lubrication. On the other hand, viscosity thermal susceptibility of HOSO samples increases with

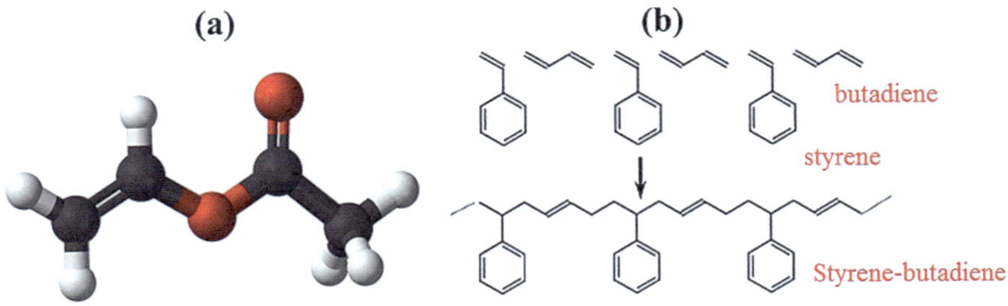

Fig. 4.2 Chemical structure of (**a**) vinyl acetate and (**b**) styrene-butadiene-styrene

EVA or SBS concentration. Consequently, plenty of research activities related to the use of additives in vegetable oils, to minimize these negative properties, are required [18, 19].

To develop the environmentally friendly lubricant, HOSO was blended with different concentrations from 1 to 5% (w/w) with either EVA copolymer, used to develop the rheology of crude oil [20], or SBS copolymer. At an earlier stage, there was instability noticed through visualization of a cloudy sample first, and phase separation occurs later on. Table 4.4 shows the kinematic viscosity of the different blends, as well as the viscosity increments about the virgin HOSO sample, as a function of temperature and concentration (EVA or SBS), respectively. As can be observed, viscosity increments are lower at high temperatures, particularly in the highest polymer concentrations. Moreover, as was previously remarked, the higher is the polymer concentration, the higher the increment in viscosity is. Therefore, for example, at 40 °C, 1% (w/w) concentration of EVA showed an increased in kinematic viscosity of 40% (63 mm^2/s) compared to virgin oil, whereas viscosity increments of up to 543% (168 mm^2/s) were found in the highest EVA concentrations (4% w/w). The similar performance was shown by SBS-based formulations, although slightly lower viscosity increase than those found in HOSO/EVA blends was obtained. The VII additive behaves entirely differently by changing the temperature, and its graphical explanation is given in Fig. 4.3.

4.6 Polysulfide and Biodegradable Polyester as an a Extreme Pressure Additive

Marketable polysulfide and biodegradable polyesters were used as extreme pressure additives and examined in nonpolar (150N) and refined polar vegetable (SOY)-based oils of the same viscosity, by a four-ball tester by Biresaw et al. [21]. They investigated a function of additive concentration on binary blends of EP additive and base oil and also tested commercial straight oil metalworking fluid with chlorinated EP additive for other comparisons. The four-ball tester produced two sets of data: pre-weld properties with peak torque and the other one wear scar diameter

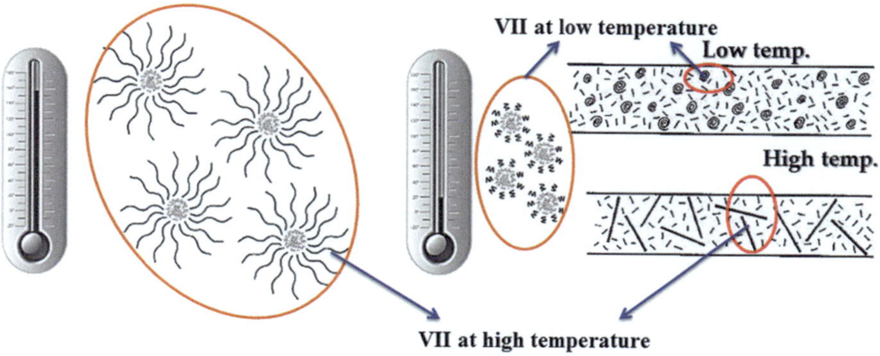

Fig. 4.3 VII additives – polymer responsible in vegetable oil at different temperatures

Table 4.5 EP property of polysulfide (PS) and biobased polyester (PE) on nonpolar paraffinic and soybean oil (base oil) [21]

EP additive, % w/w									
		Time-to-failure data from TCT EP measurements[a]					Four-ball EP weld point results (mm)		
Base oil	EP additive	0%	3%	5%	10%	20%	0%	5%	20%
150N	Polyester	3.7	–	5	12	31.7	120	160	190
150N	Polysulfide	3.7	–	4.5	10.3	13	120	590	500
SOY	Polyester	21.4	29.7	27	31.5	18.7	120	130	160
SOY	Polysulfide	21.4	27	36	39	29	120	942	981
Comm. MWF	Cl but no S P	–	–	–	–	35.3	–	–	600

[a]10 rpm or 1.2 cm/s speeds; 250 MPa contact pressure, room temperature

(WSD). The detailed comparison and EP properties of lubricants are shown in Table 4.5.

The weld point results were observed for the neat base oils and its composites:

1. The virgin base oil displayed similar WP (120 kgf) with two base oil mixtures.
2. WP increased with increasing amount of EP additive in the virgin oil.
3. Enhanced WP of both base oils while the merger of either EP additives.
4. Polysulfide showed 2.6–7.2-fold higher WP than polyester when it blends with similar virgin oils.
5. The elemental skeleton of base oil has a big effect on weld point of EP additives. Hence, at a similar ratio, polyester has a higher weld point in nonpolar oil than in polar vegetable oil, while the weld point of polysulfide was higher in the polar vegetable oil than in the nonpolar base oil. This dissimilarity was recognized of polyester being more compatible than polysulfide in the nonpolar 150N and of polysulfide being more compatible than polyester in the polar SOY (Table 4.5).

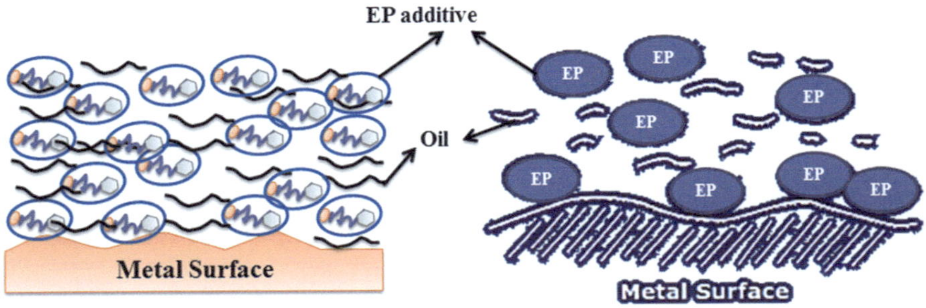

Fig. 4.4 General type of EP additives attraction with vegetable oil and metal surface

6. Composites of polysulfide in both nonpolar and polar vegetable oils displayed equivalent WP properties as commercially available MWF, the results proposing that chlorinated EP additives could be substituted with sulfurized products.

Time-to-failure (TTF) results achieved from TCT investigation are a function of base oil chemistry, EP additive chemistry, EP additive concentration, and additive-base oil relations. Consequently:

(a) Virgin vegetable oil (SOY) has a TTF result which is fivefold larger than that of neat nonpolar 150N oil.
(b) TTF results of the virgin oils are enhanced with the addition of the EP additives.
(c) TTF of nonpolar oil is improved with increasing concentration of PS or PE.
(d) TTF of polar vegetable oil improved with increasing amount of PE or PS up to 10% but diminished with increasing PE or PS amount of above 10%.

7. At identical levels, TTF of the virgin oils was dependent on the elemental skeleton of the EP additive. Thus, TTF of 150N base oil was higher with blended PE than PS, whereas for SOY, TTF was higher with blends of PS than PE.
8. TTF of EP additives is greatly reliant on chemistry of base oils. Result is that PS and PE represent the best and poorest EP additives, respectively. Comparison of WP of four-ball EP vs. TTF of TCT results showed that they are similar in some cases but different in others. The differences may be described by invoking boundary influence to TCT before the creation of in situ tribofilm. The general interactions of the EP additive with metal and oil molecules are shown in Fig. 4.4.

4.7 Methyl Methacrylate, Decyl Acrylate, and Styrene as a PPD Additive

Ghosh et al. [22] reported the alkyl methacrylate used as additives in lubricant to develop the pour point of the lubricant. In their investigation, it includes the homo- and copolymers of sunflower oil with a diverse percent of methyl methacrylate

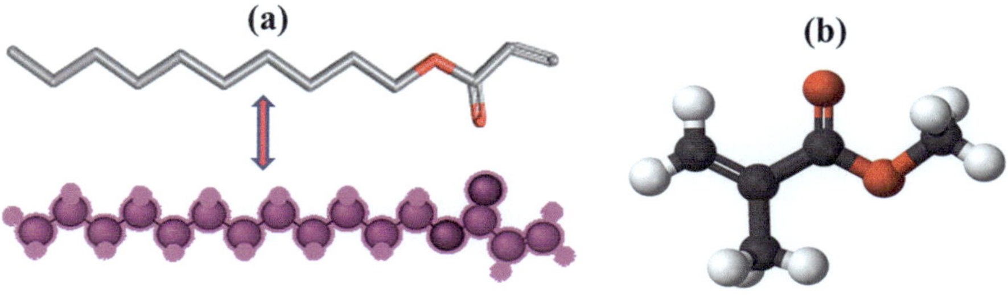

Fig. 4.5 Chemical structure of (**a**) decyl acrylate and (**b**) methyl methacrylate

(MMA), decyl acrylate (DA), and styrene and their assessments as PPD in virgin oils. The author has also observed that to provide the improved low-temperature fluidity, dispersant, solidifying. The current claim for eco-friendly technology directed us to combine the sunflower chemistry into the acrylate polymer via a copolymerization process in the expectation of producing an ideal blend of performance as well as green chemistry. Chemical structures of methyl methacrylate (MMA) and decyl acrylate (DA) are shown in Fig. 4.5.

The prepared polymers labeled as P-1, sunflower oil polymer; P-2, 5% MMA copolymer of sunflower oil; P-3, 10% MMA copolymer of sunflower oil; P-4, 5% DA copolymer of sunflower oil; P-5, 10% DA copolymer of sunflower oil; P-6, 5% styrene copolymer of sunflower oil; and P-7, 10% styrene copolymer of sunflower oil. Table 4.6 shows the pour point of lubricant doped with different percent of additives. The source and details of base oil are not mentioned by the author. The incorporation of individual monomers (such as styrene, MMA, and DA) in the sunflower oil backbone raises the thermal strength of the copolymer. Intrinsic viscosity of copolymer is less than homopolymer (except P-3), and the values obtained by the single-point determination method are higher than obtained by the graphical extrapolation method. The viscometric molecular weight of the copolymer is less than homopolymer of decyl acrylate. In all the base oils studied, PPD achievement in copolymer is always better than the homopolymer. The crystallization behavior of vegetable oil molecules with and without PPD additive is shown in Fig. 4.6.

4.8 Biodegradable Test Methods of Lubricants and Its Additives

The various methods of studying the biodegradability of lubricants were explained in the literature [23–34] (see Sects. 2.2 and 2.3 and Tables 2.1, 2.2, 2.3, and 2.4). Ghosh et al. [22] studied the biodegradability of both co- and homopolymer against

Table 4.6 Effect of methyl methacrylate (MMA), decyl acrylate (DA), and styrene in sunflower oil on pour point [22]

Pour point of additive-doped with base oil, (°C)						
Samples	Base oil		Pour point of base oil	2.5%	5%	10%
P-1	BO1	S1	−3	−6	−6	−6
		S2	−3	−3	−3	−3
	BO2	S1	−6	−9	−9	−9
		S2	−6	−9	−9	−9
P-2	BO1	S1	−3	−12	−12	−12
		S2	−3	−12	−15	−15
	BO2	S1	−6	−15	−15	−15
		S2	−6	−15	−15	−15
P-3	BO1	S1	−3	−12	−12	−12
		S2	−3	−15	−15	−12
	BO2	S1	−6	−15	−12	−12
		S2	−6	−15	−12	−12
P-4	BO1	S1	−3	−15	−12	−12
		S2	−3	−15	−12	−12
	BO2	S1	−6	−15	−12	−12
		S2	−6	−15	−12	−12
P-5	BO1	S1	−3	−15	−18	−15
		S2	−3	−15	−18	−21
	BO2	S1	−6	−18	−21	−21
		S2	−6	−18	−21	−24
P-6	BO1	S1	−3	−9	−9	−9
		S2	−3	−9	−9	−9
	BO2	S1	−6	−9	−12	−12
		S2	−6	−9	−12	−12
P-7	BO1	S1	−3	−9	−9	−9
		S2	−3	−9	−9	−9
	BO2	S1	−6	−9	−9	−9
		S2	−6	−9	−9	−9

five varieties of fungal pathogens, namely, *Colletotrichum camellias*, *Fusarium equisitae*, *Alternaria alternata*, *Colletotrichum gloeosporioides*, and *Curvularia eragrostidis*. The investigations are carried out in Petri dishes and were incubated at 37 °C for 30 days after the addition of a certain weight of polymer samples. A change of yellow confirmed the fungal growth to a blackish color. Before use, a glass apparatus and culture media were autoclaved. Culture media for fungal strains were prepared by blending appropriate proportions of potato extract, dextrose, and agar powder. The whole process was carried out in an inoculation chamber. After a

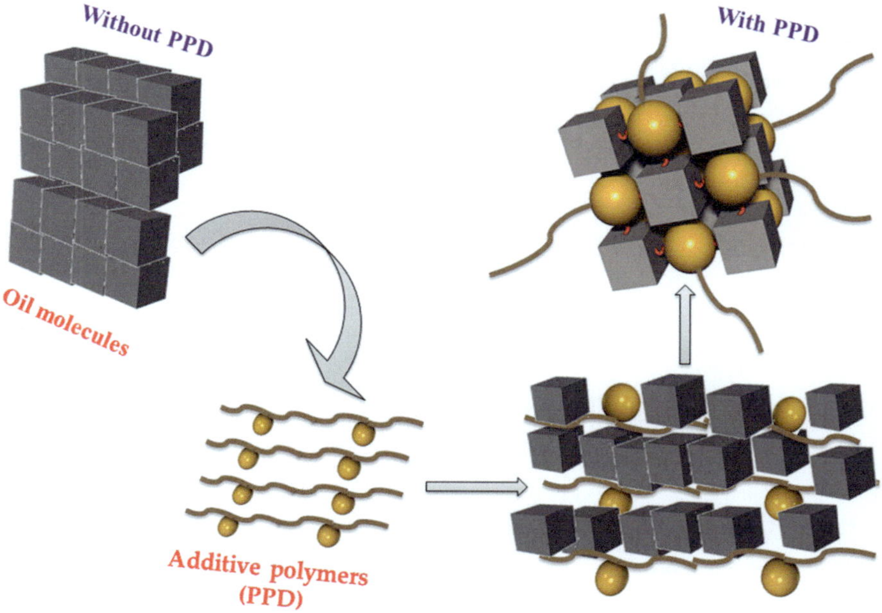

Fig. 4.6 Crystallization behavior of vegetable oil molecules with (without) PPD

test period and condition, the samples are regenerated from the condition test media and washed with suitable solvents, i.e., $CHCl_3$, purified and dried. Then the polymer was weighed and reported, and the results are given in Table 4.7.

4.9 Conclusion

Various lubrication properties, viz., pour point, viscosity index, and EP of biodegradable lubricant base oil blended and chemically treated with additives such as oleates, polyalphaolefin, diisodecyl adipate, polymethyl methacrylate, ethylene vinyl acetate, ethyl cellulose, styrene-butadiene-styrene, polysulfide, biobased polyesters, decyl acrylate, and styrene, are reviewed. Overall, due to the chemical structure and property, copolymers are having better lubrication performance than the homopolymer in the entire case reviewed in this chapter.

Table 4.7 Biodegradable test of polymer additives against five fungal pathogens [23–34]

Fungal pathogens	Polymer additive[a]	Incubation periods (day)	Initial wt (gm)	Final wt (gm)	Wt loss (gm)
Colletotrichum carnellice	P-1	30	1.5	1.5	nil
	P-2	30	1.5	1.5	nil
	P-4	30	1.5	1.5	nil
	P-6	30	1.5	1.5	nil
Fusarium equisitae	P-1	30	1.5	1.5	nil
	P-2	30	1.5	1.5	nil
	P-4	30	1.5	1.5	nil
	P-6	30	1.5	1.5	nil
Alternaria alternata	P-1	30	1.5	0.9	0.6
	P-2	30	1.5	1.1	0.1
	P-4	30	1.5	1.48	0.02
	P-6	30	1.5	1.5	nil
Colletotrichum gloeosporioides	P-1	30	1.5	1.5	nil
	P-2	30	1.5	1.5	nil
	P-4	30	1.5	1.5	nil
	P-6	30	1.5	1.5	nil
Curvularia eragrostidis	P-1	30	1.5	1.5	nil
	P-2	30	1.5	1.5	nil
	P-4	30	1.5	1.5	nil
	P-6	30	1.5	1.5	nil

[a]Composition of P-1–P-6 are given in Sect. 4.7

References

1. Maleque MA, Masjuki HH, Sapuan SM (2003) Vegetable-based biodegradable lubricating oil additives. Ind Lubr Tribol 55(3):137–143
2. Ghosh P, Pantar AV, Sarma AS (1998) Shear stability of polymers used as viscosity modifier in lubricating oil. Ind J Chem Tech 5:371
3. Abdel-Azim AAA, Nassar AM, Ahmed NS, Kafrawy AFEI, Kamal RS (2009) Pet Sci and Technol 27:20
4. Amal MN, Nahal SA (2003) J Polym Mater 52:821
5. Abdel-Azim AAA, Nassar AM, Ahmed NS, Kamal RS (2006) Preparation and evaluation of acrylate polymers as pour point depressants for lube oil. Pet Sci Technol 24(8):887–894
6. Nassar AM (2008) Synthesis and evaluation of viscosity index improvers and pour point depressant for lube oil. Pet Sci Technol 26:523–531
7. Oliveira CMF, Andrade CT, Delpech MC (1991) Properties of poly(methyl methacrylate-g-propylene oxide) in solution. Polym Bull 26:657–664
8. Delpech MC, Coutinho FMB, Habibe MES (2002) Bisphenol A-based polycarbonates: characterization of commercial samples. Polym Test 21:155–161
9. Delpech MC, Coutinho FMB, Habibe MES (2002) Viscometry study of ethylene–cyclic olefin copolymers. Polym Test 21:411–415
10. Asadauskas S, Erhan SZ (1999) Depression of pour points of vegetable oils by blending with diluents used for biodegradable lubricants. JAOCS 76(3):313–316

11. Ghosh P, Karmakar G (2014) Evaluation of sunflower oil as a multifunctional lubricating oil additive. Int J Ind Chem 5(7):1–10
12. Ghosh P, Das T, Karmakar G, Das M (2011) Evaluation of acrylate-sunflower oil copolymer as viscosity index improvers for lube oils. J Chem Pharm Res 3(3):547–556
13. Quinchia LA, Delgado MA, Franco JM, Spikes HA, Gallegos C (2012) Low-temperature flow behavior of vegetable oil-based lubricants. Ind Crop Prod 37:383–388
14. Quinchia LA, Delgado MA, Valencia C, Franco JM, Gallegos C (2009) Viscosity modification of high-oleic sunflower oil with polymeric additives for the design of new biolubricant formulations. Environ Sci Technol 43:2060–2065
15. Quinchia LA, Delgado MA, Valencia C, Franco JM, Gallegos C (2010) Viscosity modification of different vegetable oils with EVA copolymer for lubricant applications. Ind Crop Prod 32:607–612
16. Machado ALC, Lucas EF, González G (2001) Poly (ethylene-co-vinyl acetate) (EVA) as wax inhibitor of a Brazilian crude oil: oil viscosity, pour point and phase behavior of organic solutions. J Pet Sci Eng 32:159–165
17. Quinchia LA, Delgado MA, Reddyhoff T, Gallegos C, Spikes HA (2014) Tribological studies of potential vegetable oil-based lubricants containing environmentally friendly viscosity modifiers. Tribol Int 69:110–117
18. Lea CW (2002) European development of lubricants derived from renewable resources. Ind Lubric Tribol 54:268–274
19. Smith S, King R, Min D (2007) Oxidative and thermal stabilities of genetically modified high oleic sunflower oil. Food Chem 102:1208–1213
20. Qian J, Qi G, Han D, Yang S (1996) Influence of incipient chain dimension of EVA flow improver on the rheological behavior of crude oil. Fuel 75:161–163
21. Biresaw G, Asadauskas SJ, McClure TG (2012) Polysulfide and Biobased extreme pressure additive performance in vegetable vs. Paraffinic Base oils. Ind Eng Chem Res 51:262–273
22. Ghosh P, Das T, Nandi D, Karmakar G, Mandal A (2010) Synthesis and characterization of biodegradable polymer used as a pour point depressant for lubricating oil. Int J Polym Mater 59:1008–1017
23. Aluyor EO, Obahiagbon KO, Orijesu M (2009) Biodegradation of vegetable oils: a review. Afri J Biotech 8:915–920
24. Anand ON, Chhibber VK (2006) Vegetable oil derivatives: environment-friendly lubricants and fuels. J Synthetic Lubr 23:91–107
25. Bartz WJ (1998) Lubricants and the environment. Tribol Int 31:35–47
26. Goyan RL, Melley RE, Wissner PA, Ong WC (1998) Biodegradable Lubricants. Lubr Eng 54:10–17
27. Kitamura N (1993) Biodegradable lubricants. Japanese J Tribol 38:639
28. Liew Yun Hsien W (2015) Towards green lubrication in machining, springer briefs in green chemistry for sustainability. Springer, Malaysia
29. www.wiserenewables.com (2006) Wise Solutions–Renewable Lubricants–Biodegradability Primer
30. Boyde S (2002) Green lubricants. Environmental benefits and impacts of lubrication. Green Chem 4:293–307
31. Erhan SM, Kleiman R (1997) Biodegradation of estolides from monounsaturated fatty acids. J Am Oil Chem Soc 74:605
32. Bartz WJ (2006) Ecotribology: environmentally acceptable Tribological practices. Tribol Int 39:728–733
33. Nagendramma P, Kaul S (2012) Development of ecofriendly/biodegradable lubricants: an overview. Rene Sust Energ Rev 16:764–774
34. Rudnick LR (2006) Synthetics, mineral oils, and bio-based lubricants: chemistry and technology. CRC/Taylor & Francis Group, New York

Chapter 5
Nanomaterials as an Additive in Biodegradable Lubricants

Abstract The performance and effectiveness of nanoparticles as an additive in bio-based natural ester (vegetable oil) lubricants are appraised. The entire chapter deals with the lubricating properties of additives as nanoparticles, which blend with biodegradable base oil that can reduce the wear and increase load-carrying capacity of base oil remarkably, indicating the high potential of nanoparticles as lubricant additives with excellent levels of performance. The prime and significant advantages of nanoadditives do not need tribo-active elements such as phosphorus and sulfur in principle because they deposit on the rubbing surface and improve the tribological properties of the base oil, displaying excellent friction and wear reduction characteristics. The superior characteristics of nanoadditives are relatively insensitive to temperature and that tribochemical reaction is limited, compared with traditional additives.

Keywords Nanoadditives • Metal oxide • Graphene • Boron nitrides • Friction and wear • Nanoparticles • Boric acid • Tribology

5.1 Introduction

With the present developments of environmental-friendly guidelines, the contents of phosphorus and sulfur in lubricants are limited to the low level and will be possible to be inhibited for use in the future. As for the responsibility of protecting the environment, which everyone has to pay attention, "natural oils dissolve in nature without being harmful to the environment" due to their biodegradability and their organic backgrounds. Nanoparticles have been extensively focused on in many fields of science and technology, because of their unique properties entirely different from those of bulk materials or individual molecules [1]. Nanoparticles behave as lubricant additives and also have received enormous attention in the field of tribology [2, 3]. One significant property that makes the nanoparticles different from other materials is the fact that nanomaterials have an exceptionally large surface area. Due to their massive surface area, nanoparticles are extremely reactive compared to its larger form [1]. The tribological mechanism of nanoparticles as

D. Gnanasekaran, V.P. Chavidi, *Vegetable Oil based Bio-lubricants and Transformer Fluids*, Materials Forming, Machining and Tribology,
DOI 10.1007/978-981-10-4870-8_5

lubricant additives is entirely different from that of traditional antiwear and extreme pressure additives, which need to react with a rubbing surface to form protecting films [3–6]. In particular, CuO, ZnO, WS_2, TiO_2, and boron nitride (BN) nanoparticles are of extraordinary significance, as they can deposit on rubbing surfaces effectively and improve the tribological properties of the base oil, resulting in drastically reduced friction and wear as well as self-repair of worn surfaces [7–12]. The graphical representation of lubricant with/without nanoparticles as an additive in the lubricating process is shown in Fig.5.1.

Lubricants incorporated with nanoparticles are called "nanofluids" which are not simply liquid-solid mixtures. Some special requirements are essential, e.g., uniform and stable suspension, robust suspension, a negligible agglomeration of nanoparticles, and no chemical change of the fluid. However, in most of the research works, the metallic oxides are used as lubricant additives. The most important properties of nanoparticle, i.e., the tribological properties of ZnO [13–15], ZrO_2 [14–16], TiO_2 [17, 18], SiO_2 [19], CuO [14, 15, 18], and Al_2O_3 as additive in lubricants, were investigated. Some other derivatives of silicon carbide are metal nitrides such as AlN and SiN, metals like Al and Cu, and nonmetals such as graphite and carbon nanotubes, and finally in layered structure of Al + Al_2O_3, Cu + C, functionalized nanoparticles are used as an additive in the lubricants. Results meticulously showed that these nanoparticles could deposit on the rubbing surface and improve properties of the base oil (see Fig. 5.1). Moreover, a low concentration of nanoparticles is enough to develop the friction and wear performance, nano-oxides 0.5 wt% such as below 2 wt % is the optimum level. Another significant advantage of the addition of nanoparticles in the lubricant is that the filters cannot retain them on reconditioning and reclamation process. Researchers and scientists have been proposing several action mechanisms to understand nanoparticles, which has been working as additives in base fluid to reduce friction and wear. For example, Chiñas-Castillo and Spikes [20] investigated the solid colloidal nanoparticles' action mechanism in lubricants. The experimental results show that nanoparticles penetrated and formed a boundary film higher than the particle size.

In research related to literature survey [1–18], Peng et al. [19] referred the following four different action mechanisms of added nanoparticles to the base oil:

1. Nanoparticles with small size have more interaction with metal surfaces to form a surface-protective film.
2. Small nanoparticles with sphere shape have behavior to roll between the surfaces and change the sliding friction.
3. Compressive stress associated with high contact pressure can be reduced by many added nanoparticles, which bear the compressive force.
4. Physical tribofilm is formed by deposition of nanoparticles on the surface that pays advantage for the loss of mass, and this effect is called "mending effect."

A combination of these effects explained the real friction and wear of added nanoparticles on the base oil.

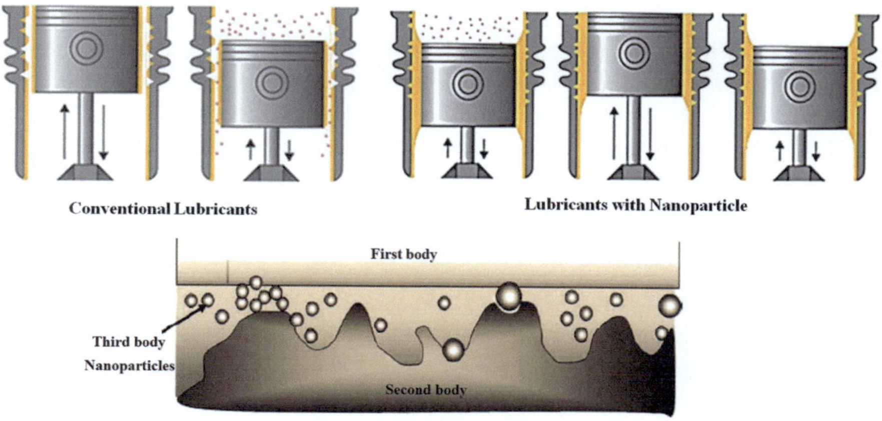

Fig. 5.1 Interaction of nanoparticles between metal surfaces

5.2 Literature Review

Prasad et al. [21] and Prasad [22] had examined the effects of various nanomaterials such as graphite, talc, MoS_2, and lead suspended in oil on the wear performance of a cast iron and zinc – aluminum alloy. Lee et al. [23] studied graphite as nanoparticle additive based on industrial gear oil to improve the lubrication properties in the disk-on-disk tribo-test. Hsu [24] discussed the important characteristics of nano-lubricants such as nonvolatile, oxidation and thermal decomposition-resistant, and self-repairing and film organization of nano-lubricant compared to conventional lubrications. The performance of grinding fluid in MQL-grinding process was improved by adding MoS_2 nanoparticles into grinding fluid [25]. Rao et al. [26] have investigated the lubricating property, turned by adding solid lubricant (graphite and boric acid) at a different particle size of 50, 100, 150, and 200 μm. The author concluded that 50 μm particle size is more efficient than other particle sizes. Huang et al. [27] indicated those graphite nano-sheets with an average diameter of 500 nm and thickness of about 15 nm in paraffin oil that formed a film on the rubbing surface not only improve the friction coefficient but also reduce wear. Through this experiment, the optimal concentrations of MoS_2 and graphite were found to be 3 wt.% for MoS_2 and 10 wt.% for graphite, respectively [28]. When the critical concentrations of MoS_2 and graphite are lesser, they did not protect the surfaces against wear. In the periodic table of elements, boron and nitrogen are the neighbors of carbon. Boron nitride (BN) was first founded in the early nineteenth century. The compound has the same number of outer shell electrons to those of graphite and diamond (carbon compound). BN has very similar properties of carbon [29] with both hexagonal and cubic crystal structures. While both forms of carbon exist as graphite and diamond in nature, both forms of BN, cubic boron nitride (cBN) and hexagonal boron nitride (hBN), are synthetic. Generally, the electrical properties are considered as the noticeable differences between carbon and BN. Due to the high electrical resistivity, BN is well known as dielectric and thermal conductive, whereas carbon is electrically and thermally conductive. The main difference in appearance is that hBN is white, while graphite is black. With the equivalent structure, hBN is expected to have the similar lubricant properties. However, the hBN is stable up to 1000 °C [29], while graphite was known to start to decompose at moderate temperatures (500 °C) [30]. The hBN has excellent chemical inertness and does not get wet by most molten metals [29]. All those advantages of hBN to make hBN a promising solid lubricant in MQL process and the cutting temperature are expected to be very high.

Boric acid has been identified as the lamellar molecular structure because it can be a potential solid lubricant. Further, it is abundant and environmentally kind, not causing health hazards to human beings. It is believed that better results can be obtained if boron crystals are nano-size; hence, they can better penetrate between the two moving parts. The resulting boric acid dispersed lubricants have the advantage of low friction properties of boric acid. The boric acid layered structure can slide over each other with relative ease and capable of reducing the friction and wear.

As both canola oil (base oil) and boric acid (additive) are environmentally friendly and renewable lubricants, Lovell et al.'s [31] study may also serve as a precursor to the improvement of multifunctional green lubricants by optimizing the particle size of the lubricant additives. They reported the comparative tribological performance of different-sized boric acid powder as additives in canola oil by pin-on-disk experiments to determine friction and wear behavior of the lubricants. Based on the experiments studied, the colloidal solutions of 20 nm in canola oil provided the better tribology performance. The colloidal solution did not degrade over time and able to continuously separate the asperities of the contacting surfaces. With respect to the lubricants with non-nanoscale particles, it was determined that the presence of smaller particles that are able to fill the surface asperities is much more important than the larger additives' ability to carry contact loads between the surfaces. In addition, the big particles may act aggressively since they are larger than the surface roughness. The enhanced frictional performance by MoS_2 submicron scale/microscale mixtures in paraffin oil. Figure 5.2 shows the chemical structure and H-bonds between the boric acid molecules.

Düzcükoğlu et al. [32] reported that initially in the combined lubricant of canola oil with boric acid, the friction coefficient was high; when boric acid particles spread uniformly in the oil, as a result the friction coefficient decreased rapidly. The obtained combination reduces the damages caused by physically covering surfaces that contact one another with a firm film due to the diamond-like hardness of boron crystals. Dependent on the size of the particle, the amount of solid particles can be blended and dispersed in the oil. The smaller the size of the nanoparticle, the greater will be the amounts of particles that can be suspended in the oil. The lubricants can, also, have other additives, which are added to improve the fundamental properties of lubricants even further.

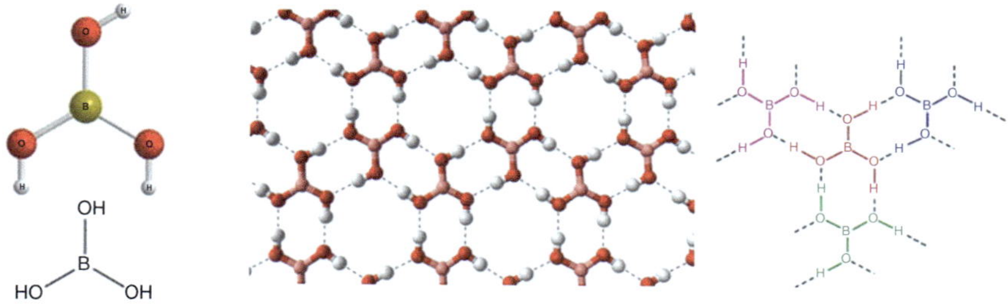

Chemical structure of boric acid **H-Bond between boric acid molecules**

Fig. 5.2 The chemical structure (**a**) and H-bonds between the molecules (**b**)

5.3 Research Scenario

As the literature cited in the previous chapters (3 & 4), the entire lubricant characteristics could differ by additive nature. The additives are of different types, such as antioxidants, detergents, dispersants, extreme pressure (EP), antiwear (AW), etc. The EP and AW are very essential at severe frictional conditions. Some of the traditional EP and AW additives are sulfur, chlorine, and phosphorus; they cover metal surfaces by forming collective layers of sulfides, chlorides, or phosphides to prevent severe wear and seizure [33–35]. The hazard chemicals such as chlorine and phosphorus have been limited for eco-friendly additives, and hence evolving new additives that pollute less or nil is the goal of future researchers. Due to the outstanding tribological and ecological properties, nanoparticles have been observed as excellent candidates to conventional AW and EP additives. The pattern of an EP additive on the lubricants is shown in Fig. 5.3, and general type of EP additive attraction with vegetable oil and metal surface and the bulk oil is shown in Fig. 4.4. In modern years, plenty of research works have been carried out on the applications of nanoparticles in the field of lubricants. Actually, tribology properties are dependent on the characteristics of nanoparticles such as size, shape, and concentration. The author [36] prepared amorphous lanthanum borate with nano-size (20–40 nm), which were blended with mineral base oil. The B_2O_3 and FeB were formed on the wear, and scar metal surface is given better wear resistance. Based on the concept of enhanced lubrication property of the vegetable oil with suitable nanoparticles, this chapter has been divided into four different parts in respective nature of nanoparticles.

5.3.1 CuO and ZnO Nanoparticles as an Additive

Plenty of literature has been reported on the addition of nanoparticles into the lubricant; all of them had focused on mineral and synthetic base oil. Very few research papers have been published on the lubrication performance of vegetable oil improved

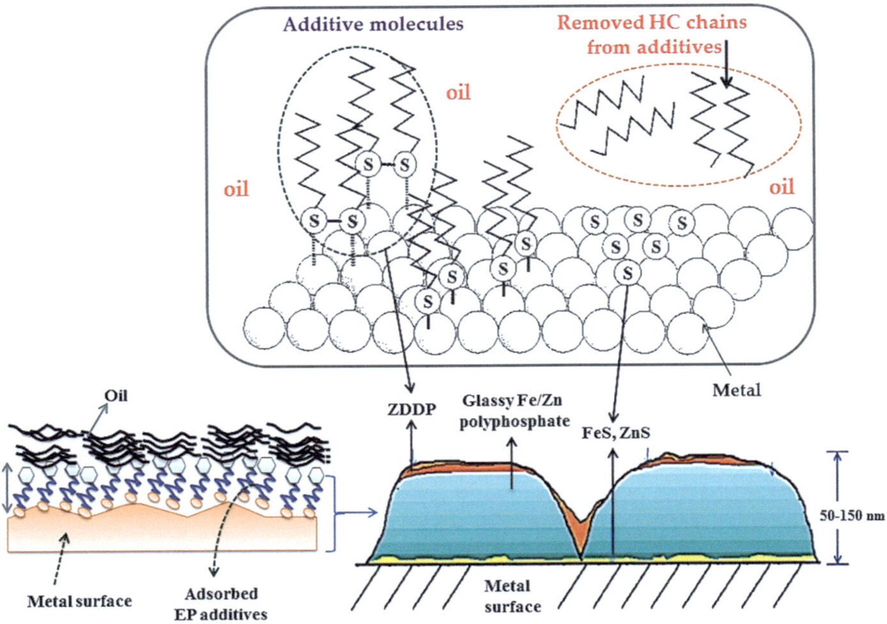

Fig. 5.3 The interaction of EP additive with metal and bulk oil

by nanoparticles. For example, Alvas et al. and Trajano et al. [37] studied and reported the antiwear and extreme pressure property of vegetable oil with CuO and ZnO as a nanoparticle using an HFRR (high-frequency reciprocating rig) equipment and SEM/EDS analysis of worn surfaces. Mineral oil and PAO without/with additive were used as a reference to analyze the influence of vegetable oil lubricant and nanoparticles on the wear of metallic surfaces. Figure 5.4 shows the general chemical structure of the metal oxide skeleton.

5.3.1.1 Friction and Wear Behavior

The tribology performance of different types of base oil with and without additives was evaluated [37, 38] and discussed in Table 5.1. The mineral and synthetic oils show the high friction coefficient without additives. Then, with the presence of nanoparticles of CuO and ZnO, the friction coefficient significantly decreases when compared with mineral and synthetic oils without additives (nanoparticle). But, this performance will not be observed in virgin natural ester fluids, in which the addition of nanoparticles showed a slight increase in the friction coefficient. The nanoparticles did not act as an antiwear additive in such case. Thus, the performance of EP additives (CuO and ZnO) is dependent on the base oil. For instance, in the case of mineral oil, the good result was detected with ZnO as additive, whereas the synthetic oil shows better synergism with CuO as additive. Probably this observation is

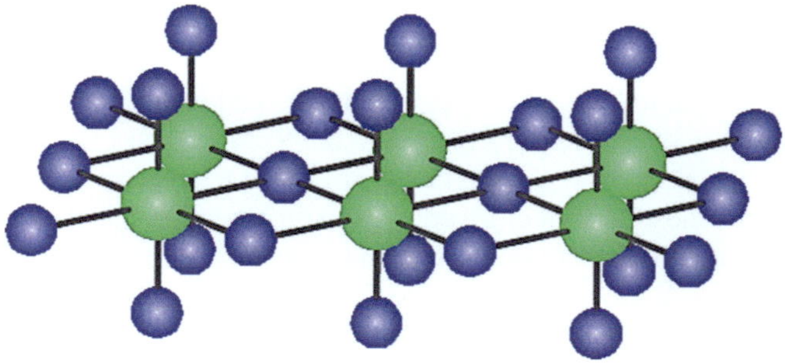

Fig. 5.4 Schematic representation of metal oxide skeleton

Table 5.1 Mean friction coefficient and standard deviation of different types of oil with/without nanoparticle [37]

Lubricants	Friction coefficient	Standard deviation
Mineral	0.104	0.00707
Mineral + CuO	0.113	0.00078
Mineral + ZnO	0.099	0.00354
Synthetic (PAO)	0.108	0.00283
PAO + CuO	0.084	0.00283
PAO + ZnO	0.096	0.00778
Sunflower	0.051	0.00071
Sunflower + CuO	0.061	0.00071
Sunflower + ZnO	0.060	0.00071
Soybean	0.053	0.00141
Soybean + CuO	0.057	0.00212
Soybean + ZnO	0.062	0.00354

due to nature of boundary lubrication, in regime chemical interactions between the lubricant and metal surface; the lubricant viscosity has little or no effect on friction and wear [38]. Only for mineral and synthetic fluids, the important characteristic of improving lubrication was nanoparticles' interaction with the metal surface. However, due to chemistry nature and its polarity of vegetable oil as a lubricant, it encourages the adsorption on the metal resulting in reduction of wear and friction, developing a thin layer for better metal-to-metal separation. During the addition of nanoparticles, they presented the behavior of the third body, increasing the friction coefficient (see Fig. 5.1). Also, according to Chiñas and Spikes [20], nanoparticles penetrate and then deposit on it because they are smaller or similar in size to lubricant film thicknesses. On the other hand, in some cases the nanoparticles have a harmful effect, increasing either friction or wear, as observed in vegetable oils with nanoparticle combination.

5.3.1.2 Worn Surface Analysis

The author [39] studied the surface morphologies and element distribution of the worn disk surfaces by SEM with EDS. From the SEM analysis, the antiwearability of blends of nanoparticle with vegetable oil can be evaluated. The experimental results showed almost no sign of severe wear (smoother and flatter) for virgin vegetable oils than the mineral and synthetic oils. Conversely, the addition of CuO and ZnO to vegetable oils observes the marks of abrasive wear against sliding direction. Hence, nanoparticles of CuO and ZnO do not have a worthy antiwearability while combined with vegetable fluids. The wear scar diameter (WSD) experiments confirmed ZnO showed little scratches with more wear while comparing CuO and ZnO effects on both sunflower- and soybean-epoxidized oils. It is understood that the vegetable oil with ZnO blends has higher WSD; at the same time epoxidized vegetable oils appear to have lower WSD. The same experiments, i.e., worn surface tests, for mineral and pure synthetic oil were carried out, and the results found that a little rough with grooves are observed when compared with vegetable oils. The metal oxide (ZnO) shows excellent behavior in friction and wear reduction when blended with mineral oil. As a result, the worn surfaces are smooth with only slight signs of wear, when CuO and ZnO were added. Hence, it is concluded that the CuO and ZnO have different performance depending on the base oil. The authors [14] established that CuO nanoparticles with synthetic oil (PAO) revealed the best EP performance. The smaller the WSD, the better behavior was observed in mineral oil with ZnO blends. The antiwear behavior of the oxide nanoparticles completely depends on the base oil nature, i.e., mineral or synthetic or vegetable oil. The oxidized vegetable oil such as sunflower and soybean oils does not display good antiwearability; this is due to the influence of the chemical nature of vegetable oil on film formation because of polar groups that adhere on the metal surface (see Sect. 1.3 and Fig. 1.3). Therefore, the nanoparticles will involve third body behavior, hence increasing the friction in this case (Fig.5.1). The synthetic oil (PAO) has excellent improved tribological performance in the presence of CuO. As per the literature survey on this theme, the antiwear mechanism is ascribed that the settlement of nanoparticle on the metal surfaces and concurrent formation of physical films could reduce the friction and wear.

According to Xue et al. [40], the tribological properties of the virgin oil can be improved by deposition of inorganic and organic nanoparticle on the metal surface. Xue et al. [40] and Dong et al. [41] have reported and also noted that nanoparticle has wear reduction as well as tremendous adhere properties even at less than 2 wt.%. The biolubricant prepared from sunflower oil and CuO and ZnO (additive) showed a similar performance of film formation around 20%, while virgin oil exhibited superior behavior about 90% after 2000 s of wear test. Similar behavior has not been observed with soybean virgin oil. The percentage of film construction was decreased drastically, while ZnO was added to a base oil, and hence biolubricant (soybean) with CuO could not retain the film throughout the test. This behavior is due to the polarity of sunflower that is higher than soybean, resulting in higher adsorption on the metal surface. According to Hutching [42], vegetable oil has a molecular structure with boundary lubrication properties (like oleic acid). In addition to friction coefficient analysis, the WSD of all the oils was evaluated to analyze the wear. The addition of CuO and ZnO on soybean increases the wear scar diameter. On the other hand, there is slight increase observed in the case of sunflower with the nanoparticle. The high wear was found in two cases: (i) the soybean lubricant with ZnO nanoparticles (248 mm) and (ii) sunflower lubricant with CuO nanoparticles (210 mm). The less wear was found in (i) epoxidized sunflower oils with additives and (ii) sunflower biolubricant without additives with approximately 180 mm. Based on literature [40–42] and the performance of additive in biolubricant, the following conclusion has been proposed:

- The excellent viscosity index properties of epoxidized sunflower and soybean oils can be used for the application in wide temperature range.
- Epoxidized sunflower oil showed good performance in boundary conditions, decreasing friction coefficient and improving film formation on the metal surface.
- As expected by literature, CuO and ZnO added to the epoxidized oil do not exhibit wear and friction reduction. It may be due to the outstanding adsorption capacity of metal oxide on the metal surface (see Fig. 1.3). Therefore, the nanoparticle deposition on the surface is hindered, and they act as a third body increasing the wear (Fig. 5.1).

5.3.2 *Boron Nitride Nanoparticles as an Additive*

Boron nitride has an isoelectronic structure with that of carbon compounds, i.e., graphite. Boron nitride exists as different forms, out of which hexagonal boron nitride (hBN) is having a layered structure having a network of $(BN)_3$ rings which are a suitable property to behave as lubricant additives (see Sect. 5.2). Figure 5.5 illustrates the different forms of BNs and suitable forms of BN for lubricant additives. Boron nitride is an insulator unlike graphite as it is having covalent interlayer bonding between boron and nitride atoms. It can localize the free electrons and makes them unavailable for electrical conductivity. The hexagonal ring network is bonded by weak van der Waals forces which permit the layers to slide over each other easily (Fig. 5.5). Those are the very important properties that help to act as the additive in lubricants. Generally, BN has good high-temperature resistance, chemical inertness, environmental compatibility, and is nontoxic. Up to now, numerous researchers have been publishing a study on the lubricity of BN against variety application. However, some of the early research reported on tribological properties of cubic, amorphous, and hexagonal boron nitride films [43] and who pointed that BN film helped to decrease the friction force and that friction coefficient was a function of average load. The interesting behavior of one form of hBN, when added to lubricating oil, and the effect of hBN concentration on friction and wear as well,

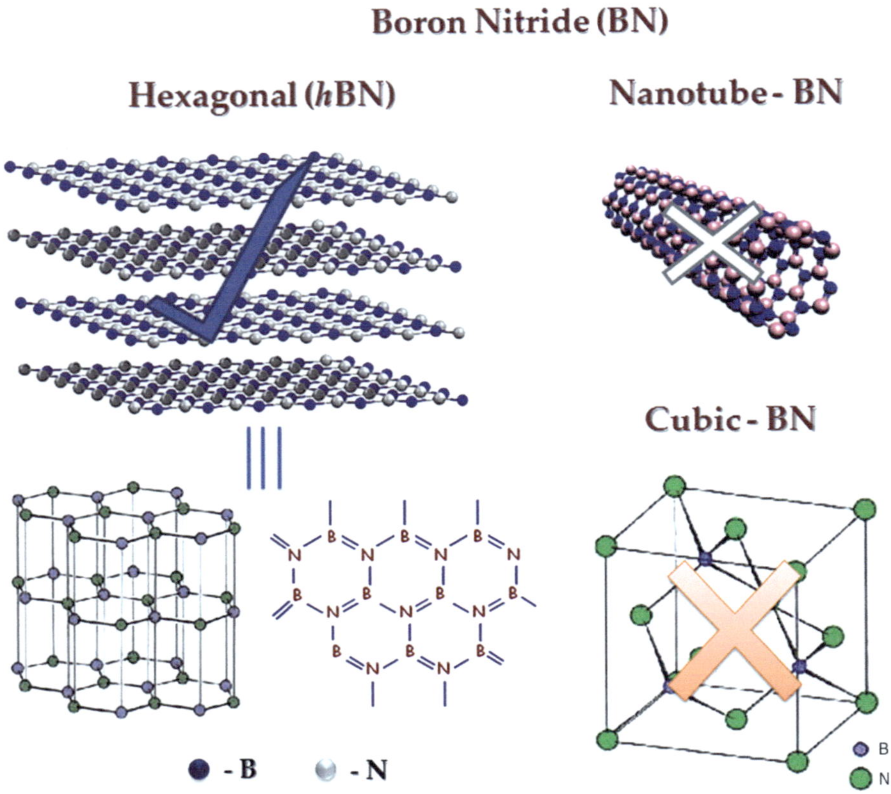

Fig. 5.5 Chemical structure: different forms of nano-boron nitride

was conducted at Kimura et al. [44] in a ring-on-roller tribometer. In sliding of bearing steel vs. itself, Kimura et al. [44] showed that addition of BN as little as 1 wt. % resulted in a substantial reduction in wear. With higher concentration, the wear is reduced, but the friction coefficient is slightly increased. Based on this research literature, BN was considered as a potential lubricating oil additive for certain applications. The author [45] tested and set the amplitude of 3 mm while carrying the load at 1, 5, and 10 N and the sliding speed at 0.25, 1.0, 2.5, and 4.0 cm/s. The mixtures include vegetable oil mixed with 0.1 and 0.5 wt% of hBN and 0.1 wt% and 1.0 wt% of graphene (xGnP) grades, M5, C300, and C750, respectively. Both xGnP and hBN have enhanced lubricants and significantly reduced the friction. The resistance behavior of the hBN/oil mixtures did not change compared to the xGnP/oil mixtures used in Park et al. [45]. The friction measurements of two distinct concentrations of xGnP (M5) and hBN were similar. These discoveries clearly indicated the greater performance of hBN nano-platelet mixtures over xGnP mixtures. While the optimum concentration of xGnP mixture was found to be 0.1w%, 0.5 wt% hBN mixture shows superior performance regarding flank wear. In general, the additions of nano-platelets help to reduce the central wear except for 0.1 wt% hBN and 0.1 wt% xGnP

Table 5.2 Specific lubricant properties of nanoparticles [61]

Properties	CuO	WS_2	TiO_2
Purity (%)	99	99.9	99.5
Size range (nm)	40–70	40–80	30–50
Color	Black	Gray	White
Morphology	Nearly spherical	Nearly spherical	Nearly spherical
Bulk density (kg/m^3)	0.76×10^3	0.25×10^3	0.42×10^3

C750. The motive may be due to the debris or other hard phases impacting the central wear region, which results in the damages by microfractures. Once the fracture occurred, the formation of wear will speed up, which will lead to a wide variety in wear measurement. Between two concentrations of hBN mixtures, both central and flank wear were reduced to a higher concentration (0.5 wt%) of hBN. The results of tribometer test allow the author to identify the advantage of using hBN platelets in reducing flank and central wear despite the fact that the friction could not change from those of xGnP. However, the author found that the highest concentration of hBN was better, and findings have been corroborated by the experimental results from the tribometer test.

5.3.3 CuO, WS_2, and TiO_2 Nanoparticles as an Additive

In the recent year, there is abundance of interesting research studies confirming that the effect of dispersed nanoparticle in lubricating oil displayed a better friction reduction and wear resistance between two metal-pairing surfaces [18, 46–52]. It is proven from the current literature that the tribological characteristics of numerous nanoparticles dispersed in base oil showed reduced friction and wore reduction between the rubbing surfaces [53–55]. Furthermore, the author found that some interesting property increases the viscosity of base oil with the addition of nanoparticles [56–60]. Baskar et al. [61] had focused on the evaluation of the tribological behavior and viscosity of nano-incorporated biolubricant using a four-ball tribometer and viscometer, respectively. They described the tribological behavior of CuO, WS_2, and TiO_2 nanoparticles as an antiwear additive blended with a chemically modified rapeseed oil (CMRO) by four-ball tester. The important properties of the nanoparticles as an additive in these tribological studies are explained in Table 5.2. The variation of viscosity of several nano-based biolubricants on temperature is calculated by ASTM D 445, and results compared with petroleum-based lubricant (SAE20W40) are shown in Table 5.3. The test results exhibited that CMRO with nano-CuO has better tribological characteristics, smoother wear scar, and higher viscosity compared to synthetic lubricant and other nano-based biolubricants. The following three conclusions have been drawn for the tests performed on a four-ball

Table 5.3 Various lubricating properties of lubricants doped with different nanoparticles [61]

Properties	*SAE20W40*	*CMRO+ CuO*	*CMRO + WS_2*	*CMRO + TiO_2*
Temperature, °C	K. viscosity, cSt			
40	123.4	128.7	128.3	127.9
50	100.1	102.3	101.9	101.2
60	80.7	83.1	82.9	82.4
70	62.2	63.9	63.3	62.9
80	43.8	45.4	45.1	44.9
90	30.1	31	30.9	30.6
100	15.2	15.6	15.3	15.0
Properties, test methods	*SAE20W40*	*CMRO + CuO*	*CMRO + WS_2*	*CMRO + TiO_2*
Pour point, ASTM D97, °C	−21	−16	−15	−15
Flash point, ASTM D92, °C	250	242	220	239
Viscosity index, ASTM D2270	133	185	179	170
Specific gravity, g/ml, @15 °C, ASTM D287	0.87	0.89	0.88	0.88
Wear scar diameter (mm), ASTM D4172	0.5849	0.3546	0.3749	0.3847
Coefficient of friction	0.10	0.07	0.08	0.07

tribometer and viscometer with synthetic lubricant and CMRO having different nanoparticles:

1. The frictional coefficient of CMRO incorporated with nano-CuO is lesser than synthetic lubricant and CMRO having nano-WS_2 and nano-TiO_2, respectively.
2. The worn wear scar value of nano-CuO embedded with CMRO is reduced than synthetic lubricant and CMRO containing nano-WS_2 and nano-TiO_2, respectively.
3. The viscosity of CMRO containing nano-CuO is higher than synthetic lubricant and CMRO containing nano-WS_2 and nano-TiO_2, respectively.

5.3.4 *Nanoparticles of Graphene Platelets (NGPs) as an Additive*

Zen and Rashmi [62] had chosen the NGP as an additive to check the tribological performances of two different biolubricants; they are:

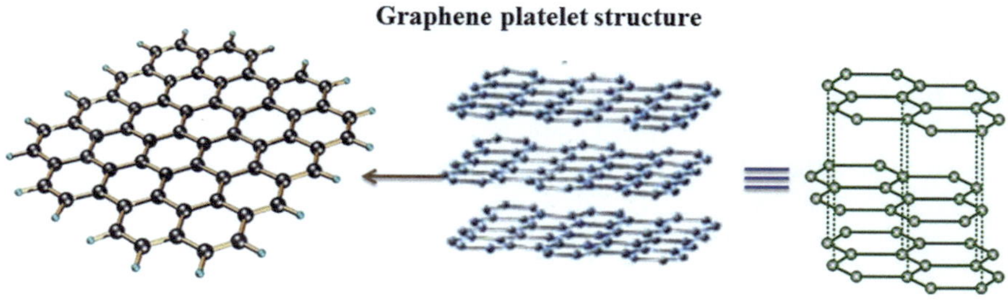

Fig. 5.6 Graphene platelet structure

(a) Trimethylolpropane (TMP) ester of palm oil
(b) A mixture of 20% TMP with 80% palm cooking oil

NGPs are distributed in each of the above ratio of 0.01 to 0.1% by weight based on the total weight of the oil and graphene combined. Followed by the addition of the desired amount of NGPs, the lubricant turns into a grease. Results showed that the incorporation of NGPs as an additive leads to the reduction of the coefficient of friction (by four-ball machine) in biolubricants, achieving a maximum reduction of 10% in first case (a) and a remarkable 17.9% decrease in the mixture (b). However, at higher temperature the system got a failure and would be due to the negative effect of graphene to form a protective film over metal surfaces and the propensity to agglomerate at higher concentrations of NGPs. Figure 5.6 illustrates the graphene platelet structure.

The tribology property was obtained from the four-ball results; it would be detected that the addition of NGP to palm oil-based TMP esters significantly decreased the friction coefficient. However, at higher temperatures, the nanobiolubricants are poorer than the blank TMP esters. The nano-lubricant dispersion produces the biggest improvement in lubricity with 0.05 wt% of NGP, roughly reducing the friction coefficient by 9.12%. Whereas at 60 ° C, more reduction in friction coefficient is influenced by the natural ester with NGP 0.1 wt%, decreasing the resistance constant by 10%. The above statement claims that the each and every single NGP layer can firmly adhere to metal surfaces, therefore producing lubricating film [63]. The addition of NGPs to blends (cooking oil and TMP) observed the same tendency as NGP with TMP, where an essential decrease is observed in the friction coefficient. The sample with 0.1 wt% of NGP brings forth the biggest reduction (17.9%) in friction coefficient at 40 °C, and at 60 ° C there is a decrease for the same sample drops to about 2%. But, the difference becomes insignificant at higher temperatures.

At higher temperatures, NGP's additive loses its lubrication property due to the damaging of the shielding film formed by the NGP at that temperature. Furthermore, at higher temperatures the single NGP layer is not capable of strongly adhering to metal surfaces. It is also indicated that wear and friction coefficient properties drop due to a significant growth in cluster formation at higher NGP concentrations [64]. To solve this problem, the addition of surfactants to the lubricant that will collapse the agglomerations occurs. The role of NGPs as a useful additive is connected to dropping the friction coefficient of lubricants, attaining a maximum reduction of 10% in TMP and an impressive 17.9% reduction in TMP with palm oil mixture. Over different temperatures tested, only the mixed blend has a better distribution of the friction coefficient.

5.4 Conclusion

Lubricating oils are commonly used to reduce the friction, prevent the wear, and carry loads needed to keep the mechanisms at better efficiency. The biolubricants from vegetable oils are naturally suitable to be used as lubricant base oils. With the cooperation of chemical modification methods to enhance its lubricating properties, the vegetable oils are functioning as well as the mineral and synthetic oils or better. The review of literature survey shows that the mechanisms of friction reduction and antiwear of nanoparticles in biodegradable lubricant have been reported as a colloidal effect, protective film, and third body. Nanoparticles blended into biodegradable lubricating oils could improve the properties of extreme pressure, antiwear, and friction reduction. Based on the works of literature cited in this chapter, which explains the selective addition of additives is crucial to increase its stability and provide the vegetable oils to work under the widest range of temperature and pressure. With the outstanding performance, vegetable oil will be a better choice to use for its nontoxic and biodegradable advantages, yet it still offers space for improvement and usefulness.

References

1. (a) Cushing BL, Kolesnichenko VL, O'Connor CJ (2004) Recent advances in the liquid-phase syntheses of inorganic nanoparticles. Chem Rev 104(9):3893–3946. (b) Gnanasekaran D, Madhavan K, Reddy BSR (2009) Developments of polyhedral oligomeric silsesquioxanes (POSS), POSS nanocomposites and their applications: a review. J Sci Ind Res 68:437–464
2. Bakunin VN, Suslov AY, Kuzmina GN, Parenago OP, Topchiev AV (2004) Synthesis and application of inorganic nanoparticles as lubricant components – a review. J Nanopart Res 6(2/3):273–284
3. Zhang ZJ, Zhang J, Xue QJ (1994) Synthesis and characterization of a molybdenum disulfide nanocluster. J Phys Chem 98(49):12973–12977
4. Margielewski L, Stanecka-Badura R, Plaza S, Kozlowski W (2012) Preparation and tribological properties of zinc dialkyl(aryl)ethoxy dithiophosphates. Industrial Lubrication and Tribology 64(4):187–195
5. Lee K, Hwang Y, Cheong S, Choi Y, Kwon L, Lee J, Kim SH (2009) Understanding the role of nanoparticles in Nano-oil lubrication. Tribol Lett 35(2):127–131
6. Tarasov S, Belyaev S (2004) Alloying contact zones by metallic nanopowders in sliding wear. Wear 257(5/6):523–530
7. Tarasov S, Kolubaev A, Belyaev S, Lerner M, Tepper F (2002) Study of friction reduction by nanocopper additives to motor oil. Wear 252(1/2):63–69

8. Yidong Z (2012) Focus on self-repairing performance and mechanism of cu-DDP additive in base lubricating oil. Industrial Lubrication and Tribology 64(1):11–15
9. Zhang M, Wang X, Liu W, Fu X (2009) Performance and anti-wear mechanism of cu nanoparticles as lubricating oil additives. Industrial Lubrication and Tribology 61(6):311–318
10. Zhang YD, Yan JS, LG Y, Zhang PY (2009) Effect of Nano-cu lubrication additive on the contact fatigue behavior of steel. Tribol Lett 37(2):203–207
11. Zhou J, Wu Z, Zhang Z, Liu W, Xue Q (2000) Tribological behavior and lubricating mechanism of cu nanoparticles in oil. Tribol Lett 8(4):213–218
12. Zhou J, Yang J, Zhang Z, Liu W, Xue Q (1999) Study on structure and tribological properties of surface-modified cu nanoparticles. Mater Res Bull 34(9):1361–1367
13. Hernandez Battez A, Fernandez Rico JE, Navas Arias A, Viesca Rodriguez JL, Chou Rodriguez R, Diaz-Fernandez JM (2006) The tribological behavior of ZnO nanoparticles as an additive to PAO6. Wear 261:256–263
14. Hernandez Battez A, González R, Felgueroso D, Fernández JE, Rocio Fernández MR, García MA (2007) Wear prevention behavior of nanoparticle suspension under extreme pressure conditions. Wear 263:1568–5747
15. Hernandez Battez A, González R, Viesca JL, Fernández JE, Dias Fernández JM, Machado A (2008) CuO, ZrO2 and ZnO nanoparticles as antiwear additive in oil lubricants. Wear 265:422–428
16. Ma S, Zheng S, Cao D, Guo H (2010) Anti-wear and friction performance of ZrO2 nanoparticles as lubricant additive. Particuology 8:468–472
17. YY W, Kao MJ (2011) Using TiO_2 nanofluid additive for engine lubrication oil. Industrial Lubrication and Tribology 63(6):440–445
18. YY W, Tsui WC, Liu TC (2007) Experimental analysis of tribological properties of lubricating oils with nanoparticles additives. Wear 262(7):819–825
19. Peng DX, Kang Y, Hwang RM, Shyr SS, Chang YP (2009) Tribological properties of diamond and SiO2 nanoparticles added in paraffin. Tribol Int 42:911–917
20. Chiñas-Castillo F, Spikes HA (2000) The behavior of colloidal solid particles in elastohydro dynamic contacts. Tribol Trans 43(3):387–394
21. Prasad BK, Rathod S, Yadav MS, Modi OP (2010) Effects of some solid lubricants suspended in oil toward controlling the wear performance of a cast iron. J Tribology 132(4):1–9
22. Prasad BK (2005) Effectiveness of an externally added solid lubricant on the sliding wear response of a zinc–Aluminium alloy, its composite and cast iron. Tribol Lett 18:135–143
23. Lee CG, Hwang YJ, Choi YM, Lee JK, Choi C, JM O (2009) A study on the Tribological characteristics of graphite Nano lubricants. Int J Precis Eng Manuf 10(1):85–90
24. Hsu MS (2004) Nano-lubrication: concept and design. Tribol Int 37:537–545
25. Shen B, Malshe AP, Kalita P, Shih AJ (2008) Performance of novel MoS2 nanoparticles based grinding fluids in minimum quantity lubrication grinding. Transaction of NAMRI/SME 36:357–364
26. Rao DN, Krishna PV (2008) The influence of solid lubricant particle size on machining parameters in turning. Int J Mach Tool Manu 48:107–111
27. Huang HD, JP T, Gan LP, Li CZ (2006) An investigation of tribological properties of graphite nanosheets as an oil additive. Wear 261:140–144
28. Bartz WJ (1971) Solid lubricant additives - the effect of concentration and other additives on anti-wear performance. Wear 17(5–6):421–432
29. Haubner R, Wilhelm M, Weissenbacher R, Lux B (2002) Boron nitrides-properties, Synthesis, and applications, high-performance non-oxide ceramics II. Structure & Bonding 102:1–45
30. Zhu Q, Qiu X, Ma C (1998) The oxidation resistance improvement of matrix graphite of spherical fuel elements by slip-gelation process. J Nucl Mater 254:221–225
31. Lovell MR, Kabir MA, Menezes PL, Higgs CF (2010) Influence of boric acid additive size on green lubricant performance. Phil Trans R Soc A 368:4851–4868

32. Düzcükoğlu H, Acaroğlu M (2009) Lubrication properties of vegetable oils combined with boric acid and determination of their effects on wear. Energy Sources, Part A: Recovery, Utilization, and Environmental Effects 32(3):275–285
33. (a) Choi Y, Lee C, Hwang Y, Park M, Lee J, Choi C (2009) Tribological behavior of copper nanoparticles as additives in oil. Curr Appl Phys 9:124–7. (b) Margielewski L, Stanecka-Badura R, Plaza S, Kozlowski W (2012) Preparation and tribological properties of zinc dialkyl(aryl)ethoxy dithiophosphates. Ind Lubr Tribol. 64(4):187–195
34. Yu H, Xu Y, Shi P, Xu B, Wang X, Liu Q (2008) Tribological properties and lubricating mechanisms of cu nanoparticles in lubricant. Trans Nonferrous Metals Soc China 18:636–641
35. (a) Lee K, Hwang Y, Cheong S, Choi Y, Kwon L, Lee J (2009) Understanding the role of nanoparticles in Nano-oil lubrication. Tribol Lett 35(2):127–131. (b) Nicholls MA, Do T, Norton PR, Kasrai M, Bancroft GM (2005) Review of the lubrication of metallic surfaces by zinc dialkyl-dithiophosphates. Tribol Int 38(1):15–39
36. ZS H, Dong JX, Chen GX, He JZ (2000) Preparation and tribological properties of nanoparticle lanthanum borate. Wear 243:43–47
37. (a) Alves SM, Barros BS, Trajano MF, Ribeiro KSB, Moura E (2013) Tribological behavior of vegetable oil-based lubricants with nanoparticles of oxides in boundary lubrication conditions. Tribol Int 65:28–36. (b) Trajano MF, Freitas Moura EI, Ribeiro KSB, Alves SM (2014) Study of oxide nanoparticles as additives for vegetable lubricants. Mater Res 17:1124–1128
38. Hsu SM, Gates RS (2005) Boundary lubricating films: formation and lubrication mechanism. Tribol Int 38:305–312
39. Zhang ZJ, Simionesie D, Schaschke C (2014) Graphite and hybrid nanomaterials as lubricant additives. Lubricants 2:44–65
40. Xue QJ, Liu WM, Zhang ZJ (1997) Friction and wear properties of the surface modified the TiO2 nanoparticle additive in liquid paraffin. Wear 213(1–2):29–32
41. Dong C, Cao J, Struble EJ, Lipowsky HH (1999) Mechanics of leukocyte deformation and adhesion to endothelium in shear flow. Ann Biomed Eng 27(3):298–312
42. Hutching IM (1992) Tribology: friction and wear of engineering materials. Edward Arnold, London
43. Watanabe S, Miyake S, Murakawa M (1991) Tribological properties of cubic, amorphous and hexagonal boron nitride films. Surf Coat Technol 49:406–410
44. Kimura Y, Wakabayashi T, Okada K, Wada T, Nishikawa H (1999) Boron nitride as a lubricant additive. Wear 232:199–206
45. Park KH, Ewald S, Kwon P (2011) Effect of Nano-enhanced lubricant in minimum quantity lubrication (MQL) balling milling. J Tribology 39(12):517–521
46. (a) Choi Y, Lee C, Hwang Y, Park M, Lee J, Choi C, Jung M (2009) Tribological behavior of copper nanoparticles as additives in oil. Curr Appl Phys 9(2):124–127. (b) Nguyen TK, Do I, Kwon P (2012) A tribological study of vegetable oil enhanced by nano-platelets and implication in MQL machining. Int J Precis Eng Manuf 13(7):1077–1083
47. Arumugam S, Sriram G (2014) Synthesis and characterization of rapeseed oil bio-lubricant dispersed with nano copper oxide its effect on wear and frictional behavior of piston ring cylinder liner combination. Journal of Engineering Tribology 228(11):1308–1318
48. Ingole S, Charanpahari A, Kakade A, Umare SS, Bhatt DV, Menghani J (2013) Tribological behavior of nano TiO2 as an additive in base oil. Wear 301(1–2):776–785
49. Arumugam S, Sriram G (2013) Preliminary study of Nano and Microscale TiO_2 additives on Tribological behavior of chemically modified rapeseed oil. Tribol Trans 56(5):797–805
50. Rapoport L, Leshchinsky V, Lapsker I, Volovik Y, Nepomnyashchy O, Lvovsky M, Popovitz P, Feldman BY, Tenne R (2003) Tribological properties of WS_2 nanoparticles under mixed lubrication. Wear 255(7–12):785–793
51. Greenberg R, Halperin G, Etsion I, Tenne R (2004) The effect of WS_2 nanoparticles on friction reduction in various lubrication regimes. Tribol Lett 17(4):179–186

52. Lahouij I, Dassenoy F, De KL, Martin JM, Vacher B (2011) In situ TEM observation of the behavior of an individual fullerene-like MoS_2 nanoparticle in a dynamic contact. Tribol Lett 42(2):133–140
53. Chiong Ing T, Rafiq AKM, Azli Y, Syahrullail S (2012) Tribological behavior of refined bleached and deodorized palm olein in different loads using a four-ball tribotester. Scientia Iranica B 19(6):1487–1492
54. Unal H, Sincik M, Izli N (2009) Comparison of some engineering properties of rapeseed cultivars. Ind Crop Prod 30(1):131–136
55. Rudnick LR (2006) Synthetics, mineral oils, and bio-based lubricants: chemistry and technology, 2nd edn. CRC/Taylor & Francis, Boca Raton. ISBN 9781439855379
56. Namburu PK, Kulkarni DP, Misra D, Das DK (2007) Viscosity of copper oxide nanoparticles dispersed in ethylene glycol and water mixture. Exp Thermal Fluid Sci 32(2):397–402
57. Rashin MN, Hemalatha J (2013) Viscosity studies on novel copper oxide–coconut oil nanofluid. Exp Thermal Fluid Sci 48:67–72
58. Kole M, Dey TK (2010) Viscosity of alumina nanoparticles dispersed in car engine coolant. Exp Thermal Fluid Sci 34(6):677–683
59. Yiamsawas T, Mahian O, Selim A, Suthep D, Kaewnai WS (2013) Experimental studies on the viscosity of TiO2 and Al2O3 nanoparticles suspended in a mixture of ethylene glycol and water for high-temperature applications. Appl Energy 111:40–45
60. Aladag B, Halelfadl S, Doner N, Maré N, Duret S, Estellé P (2012) Experimental investigations of the viscosity of nanofluids at low temperatures. Appl Energy 97:876–880
61. Baskar S, Sriram G, Arumugam S (2015) Experimental analysis of Tribological behavior of Nano bio-lubricants using four-ball Tribometer. Tribology in Industry 37(4):449–454
62. Arwin Goo Zen, Rashmi G W, Eureca (2013) Tribological Evaluation of Nano Graphene Platelets as an Additive to Biolubricant Base Fluid, EURECA 2013
63. Hong H, Marquis FDS, Waynick JA (2007) Carbon Nanoparticle-Containing Lubricant and Grease, US Publication No. 2007/0158609
64. Bhaumik S, Prabhu S (2013) Efficiency of Multiwalled carbon nanotubes in Nanolubricant. Golden Research Thoughts 2:11

Part III
Insulating Fluids from Vegetable Oil

Chapter 6
Vegetable Oil: An Eco-friendly Liquid Insulator

Abstract This chapter deals with the basics and chemistry of natural esters and their applications as a liquid insulator. This chapter also describes the refining and processing techniques of various vegetable oil-based natural esters. This chapter also includes detailed discussion of vegetable esters of different origin and their suitability as a liquid insulator. The historical evolution of vegetable esters as transformer insulating fluid was elaborately presented in this chapter. This chapter also distinguishes the natural esters and mineral oils as a liquid insulator. This also gives detailed discussion on current research scenario of vegetable oil-based ester fluids for the power sector.

Keywords Natural esters • Triglycerides • Palm oil • Soybean oil • Coconut oil • Vegetable oils • Biodegradable

6.1 Introduction

The energy requirement throughout the world is conquered by hydrocarbon oils for hundreds of years in many industrial applications including power sector, conveyance, and household. Petroleum-based oil is a significant dielectric material in transformer industry and has been used for more than one-and-half century. Hydrocarbon-based mineral oil usage in power equipment is seriously hazardous to the atmospheric environment especially when there are any incidents during working condition of the power equipment, viz., explosion of transformer causing a spillage of oil into earth and other water streams. With the aim of reduction in the usage of hydrocarbon-based resources and environmental issues, the alternative liquid insulator with biodegradable characteristics is attracting lot of attention for about 20 years. Hence, in this chapter detailed discussion on various aspects of vegetable-based esters for power sector. The brief content of the chapter is given in Fig 6.1.

D. Gnanasekaran, V.P. Chavidi, *Vegetable Oil based Bio-lubricants and Transformer Fluids*, Materials Forming, Machining and Tribology,
DOI 10.1007/978-981-10-4870-8_6

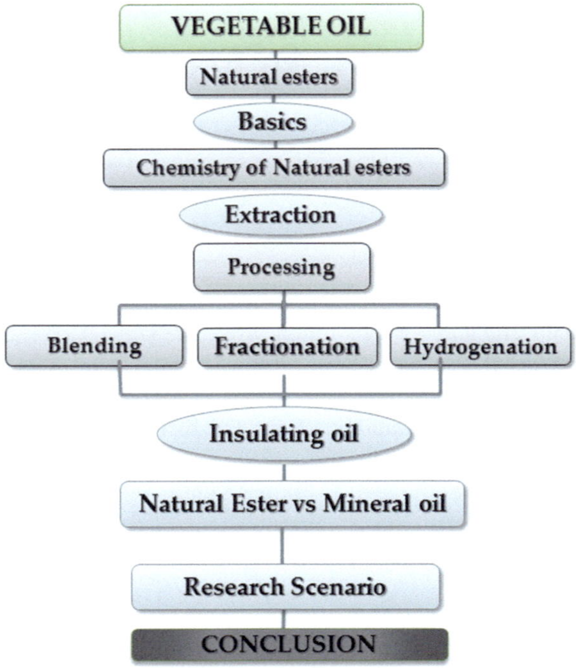

Fig. 6.1 Flow chart – content of the chapter

6.2 Natural Esters

Esters are the biological compounds synthesized by combining acids and alcohols [1, 2]. Vegetable oil esters basically consist of triglycerides, synthesized by esterification of glycerol with three different or same fatty acids and composed of linear hydrocarbon chains ended by –COOH group which consists of an even number of "C" atoms (8 to 22 in triglycerides), and the hydrocarbon chain can be saturated are mono-, di-, and tri-unsaturated. Example of a saturated fatty acid is stearic acid, and the general formula is $HOOC–(CH_2)_{16}–CH_3$, and it is expressed symbolically as C18:0, where the numbers in the molecule represent a number of carbon atoms and unsaturated bonds, respectively.

Vegetable oil produced from various plant-based seeds can be categorized by the relative quantities of fatty acids (see Table 6.1) present in it. In determining the physical properties of bio-esters, nature of fatty acid compounds of triglycerides plays a significant role. Even in small amounts, other compounds give their steadiness, color, and odor to vegetable oils. These compounds can be eliminated during the refining process depending on the individual applications. Figure 6.2 shows the general esterification process.

Table 6.1 Fatty acid content of selected plant based oils [3]

Type of vegetable oil	Palmitic (16:0)	Stearic (18:0)	Oleic (18:1)	Linoleic (18:2)	Linolenic (18:3
Safflower	6.4	2.5	17.9	73.2	–
Safflower (high-oleic content)	4.6	2.2	77.5	13.2	–
Sunflower oil	6.7	2.6	14.6	75.2	–
Sunflower oil (high-oleic content)	6.1	5.3	21.4	66.4	–
Soybean oil	3.5	4.4	80.3	10.4	–
Soybean oil (high-oleic content)	6.0	5.2	20.2	63.7	5.0
Corn oil	6.2	3.0	83.6	3.7	1.7
Cottonseed oil	10.6	2.0	26.7	59.8	0.9

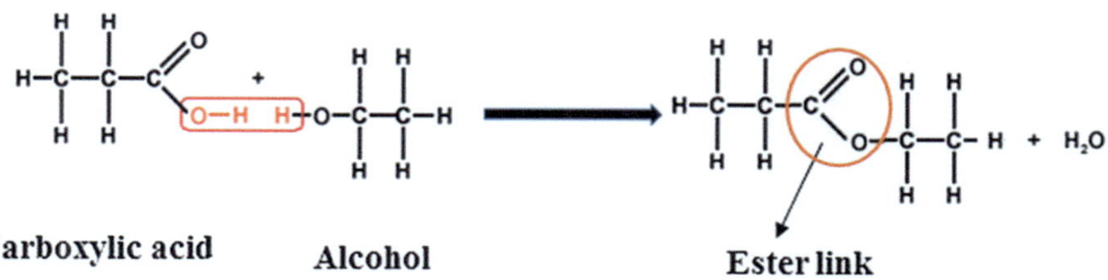

Fig. 6.2 General representation of esterification process

6.2.1 Chemistry of Natural Esters

Esters are widespread chemical complexes, and vegetable esters represents to those present in biological species. Most naturally occurring esters are triglycerides of fats and oils. Carboxylic acids and alcohols are the basic chemical compounds used for the synthesis of esters by a condensation reaction.

In organic chemistry, triglycerides are the esters derived from fatty acids and glycerols. Glycerol is a compound with three-OH functional groups. Fatty acids are carboxylic acids with a long aliphatic unbranched linear chain which is either unsaturated or saturated. Fatty acids are widely available in nature, which have a chain of an even number of carbon atoms (4 to 28) (Fig. 6.3). So, there are many triglycerides depending on various combinations of three fatty acids bonded to a glycerol backbone that depends on the source of oil (see Table 6.1).

Triglycerides (TGs) are variable in their viscosity, melting point, stability, and other critical properties. The characteristics of a TG molecule depend on fatty acids that constitute it. As presented in Fig. 6.3, the fatty acids vary by number of hydrogen and carbon atoms in the structural formula. The carbon atoms each attached to two neighboring carbons form a flexible twisting chain. Fatty acids that contain extended chains are prone to intermolecular interaction leading to increase in melting/pour point and viscosity of fatty esters. Conversely, unsaturated triglycerides

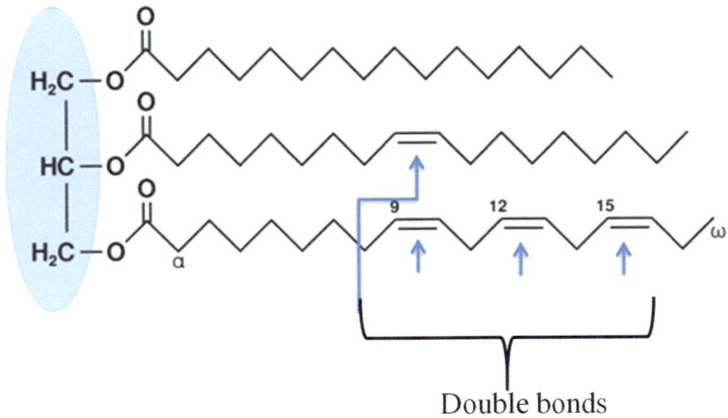

Fig. 6.3 Triglyceride (TG) with three different fatty acids (C16:0, C18:1, and C18:3)

cannot pile themselves in a closely packed arrangement as saturated TGs do; this might be due to rigid links interleaved by double bonds in the fatty acid. Hence, the unsaturated triglycerides flow and freeze less easily and are in liquid state at room temperature. For example, due to the presence of high content of saturated fatty acids, animal fats are solid state, and plant-based olive and linseed oils are highly unsaturated and are viscous liquids. Due to the presence of double bonds, unsaturated triglycerides are more prone to oxidation reaction which undergoes polymerization and degradation more easily than saturated triglycerides. A developed grade of saturation of the molecules resembles to higher oxidation stability and has high pour point and viscosity. Hence, the above considerations are helpful in deciding the best natural esters for insulating applications.

The most prevailing technology for processing vegetable oil (conversion triglycerides to fatty acid esters) is transesterification and esterification. The other possible routes for the synthesis of natural esters are emulsification, hydrotreating, dilution of the vegetable oils, and pyrolysis. The esterification process has been a widely used method to reduce the viscosity of TGs. In this process, organic group R″ of an ester exchange with the organic group R′ of an alcohol. Synthesis of natural esters from vegetable oils generally follows three steps: (i) extraction, (ii) refining, and (iii) processing and detailed discussion is as follows:

6.2.2 Extraction Technique

Esters are extracted from animal fats, plants, fruits, and seeds using various processing techniques. Various seed oils extracted and processed before they are ready to be used as transformer insulating oil are presented in Fig. 6.4.

Fig. 6.4 Vegetable seed oil from different plant origin: (**a**) coconut (**b**) soybean (**c**) palm (**d**) castor (**e**) rapeseed (**f**) sunflower oil

6.2.3 Refining Technique

Refining is a purification procedure of a material to obtain oil with required properties from crude oils through various treating techniques, viz., degumming, neutralization, bleaching, and deodorization. Bleaching is well defined as the elimination of color and oxidizing agents, gums, lather, and metals in trace amounts by mixing oil with distinct adsorbents. The adsorbents are further filtered to remove the contaminants. Deodorization is a process of removing the fatty acids, odor, and flavor and destabilizing contaminates by undergoing the oil to temperature and high vacuum environments to remove impurities by vaporization during the process; the oil remains in liquid state [4, 5]. The refining process converts the crude oil or fat into a suitable product, as shown in Fig. 6.5.

6.2.4 Processing Technique

Vegetable oils have only limited industrial applications when used in their instinctive form. Therefore, they are physically or chemically modified to improve characteristics. The widely used techniques are hydrogenation and fractionation. Among these available techniques, oil fractionation is absolutely the supreme choice due to economical viability, zero oil loss, and reversibility. Processing of vegetable oil by esterification reaction is given in Fig. 6.6.

A quite number of queries have been raised with regard to consequence of chemical treatment on quality of oils. Various techniques have been advanced, and currently prevailing processing methods have been improved in order to develop the

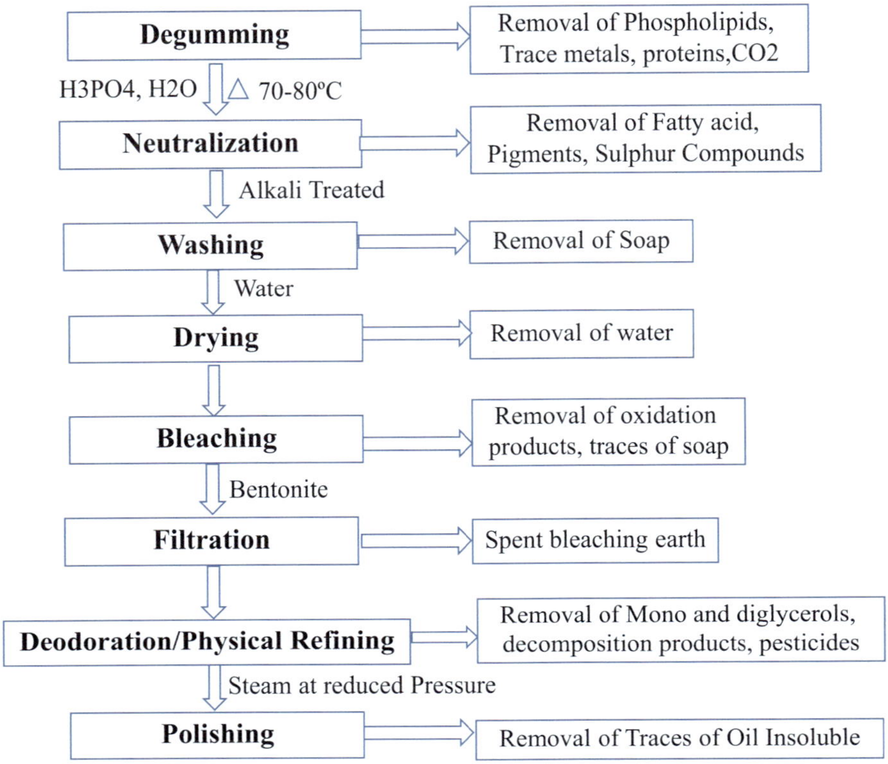

Fig. 6.5 Refining process of ester oil

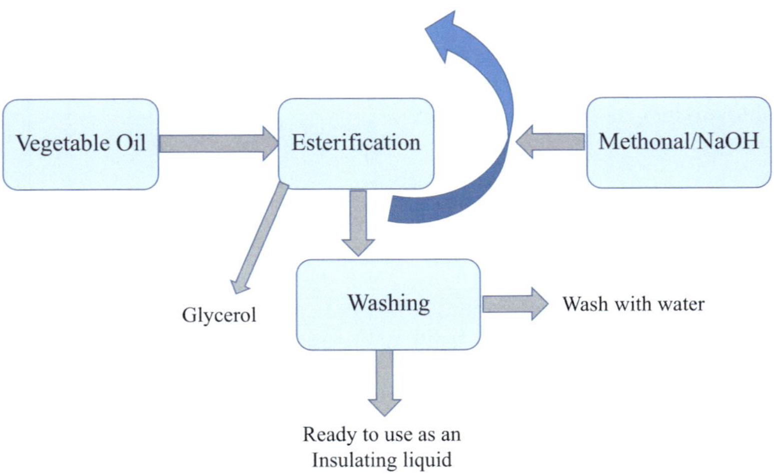

Fig. 6.6 Processing of vegetable oil by esterification reaction

quality standards. A more integrated and combined use of various treatment techniques has become indispensable. The final product of ester oil needs enhancement of some physical properties; few techniques have already been developed in this process, including blending, fractionation, hydrogenation, and esterification.

6.2.4.1 Blending Process

In an exertion to defend the environment issues, biodegradable products are suggested to be the best choice. Deployment of oils from renewable biodegradable sources can solve the environment issues. Biodegradation is an indication of the perseverance of a particular matter in the atmosphere, and it is a measure to evaluate the ecological balance of substances. Mineral oil is hydrocarbon-based nondegradable extracted from petroleum-based product and takes quite a long time to degrade. Mineral oil can cause ecological imbalance on spillage or leakage in the transformer during service and on disposal after losing its insulating properties. Toxicity, contamination of water, and waste treatment are other environmental issues restricting the usage of mineral oil. In addition to pollution, cleanup costs of leakage or spills and replacement cost are very expensive. Due to the above facts, it is a key to search for alternative to the existing mineral oil. Due to the high flash point and high breakdown voltage, vegetable oils have been reported as the better choice to replace the mineral oils as a liquid insulator and cooling agent in electrical equipment. In addition, vegetable-based insulting oils owned superior moisture absorption capacity from the solid insulation (kraft paper), hence, suggested as a feasible candidate as an alternative to existing petroleum mineral oils. However, natural esters has drawback in certain properties, viz., higher viscosity, pour point, and poor oxidation stability compared to mineral oil. Several studies have been performed and have proved various vegetable oils, viz., soybean, rapeseed, palm, sunflower oil, etc., suitable as a liquid insulator for power transformers [6]. However, studies on long-term electrical and chemical properties are lacking. Recently some researchers described that combination of natural esters and mineral oils showed superior properties and other environmental aspects [6], but no studies on their electrical properties have been performed.

Usman et al. [7] investigated various characteristics of blends of soybean and palm oil for its suitability as a liquid insulator in power equipment (transformer). Haribabu et al. [8] studied the effect of blending of the natural ester with different, triglyceride ratio and claimed that the properties obtained are sufficient to use the oil as a liquid insulator in transformers. The researchers blended high-oleic content-based natural esters with a significantly increased amount of stabilizers [8]. These attempts can result in anticipated oxidation stability requirement. Various attempts were also made by many researchers on blending of vegetable esters with less environmentally friendly synthetic ester oils to obtain a natural ester fluid meeting stability of aging characteristics of mineral oil.

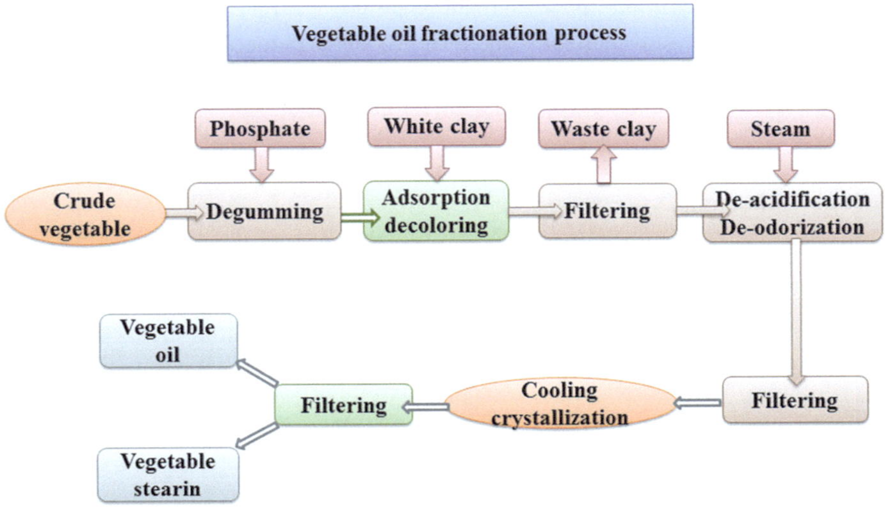

Fig. 6.7 Fractionation process of vegetable oils

6.2.4.2 Fractionation Process

Fractionation of vegetable oils denotes to the machine-driven separation process of a liquid from solid, crystallized, as components of a particular fat or oil. In the oil industry, fractionation is a process that has been known for more than ten decades. Earlier, liquid and solid portions of various oils and fats have been parted by settle-down process, using the gravitational force to separate solid phase (heavier phase) and liquid phase (lighter phase). Usually, in fractionation process, the solid phase (settled at the bottom) contains a large quantity of entrained (trapped) liquid.

The fractionation method is used as an individual method or as a combination of more complicated methods. Thus, the method is united with hydrogenation, either hydrogenation being first stage followed by fractionation to separate stearin formed or with hydrogenation applied to one of the portions produced in a fractionation technique. Fractionation route can be also combined by interesterification to randomize a fraction obtained in the process or to interesterify a mixture, a constituent a fraction produced by the above method as shown in Fig. 6.7.

6.2.4.3 Hydrogenation Process

Physical properties, viz., melting point, crystallinity will be affected by the number of double bonds present in oils and fats. Generally, double bonds reduce the oil-melting point; therefore, oil rich in unsaturated fatty acids is in liquid state, while one with a small amount of unsaturated fatty acids are in solid or semisolid state. Hydrogenation is a process of converting liquid oil to solid or semisolid state. The

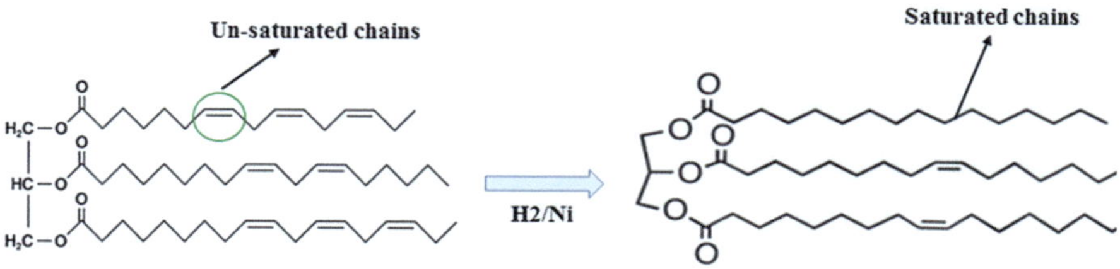

Scheme 6.1 Hydrogenation reaction in presence of nickel catalyst

addition of hydrogen atoms to the double bonds of unsaturated fatty acids (Scheme 6.1) takes place resulting to hydrogenation process. Vegetable oils are hydrogenated with H_2 gas in the presence of catalyst (nickel). If the process performed completely, all double bonds convert to saturated compounds with equal number of carbons.

6.3 Vegetable Oil as a Transformer Insulating Fluid

The rapid growth of the use of electricity probes for an advancement of power equipment with a high-level dependability and safety. Power grid connects the power plants through transmission lines and distribution lines to the end users. The main objective is providing electricity to end users in form of ready to use.

Transformer is a vital systems in power generation plants for voltage-level alteration and retaining the power supply. Transformers are used at four major stages ranging from power generation station to household and industrial applications [9], viz., (1) power plants (power is generated and raised to transmission), (2) transferring stations (alter to transmission voltage), (3) distribution substations (transmission level voltage is changed to distribution voltage), and (4) service transformers (voltage is abridged to operation level for directing into household/industries).

In power equipment, two basic insulation types are available, viz., solid and liquid. Solid insulation is made up of kraft paper, pressboard, polymer resin, and wood. Insulating paper made of unbleached softwood pulp through sulfate process is widely used as a solid insulation in power transformers. Liquid insulation of power transformer offers two main tenacities during operation: (1) dielectric material and (2) heat dissipation agent. There are various requirements for the liquid insulator to use a dielectric material in a transformer:

- As a coolant - Absorbing heat from core and windings and dissipating to exterior surface of the transformer. At higher temperatures, viscosity of oil decreases providing circulation of oil (pour point to be kept at low and capable to flow).
- As an insulator - to insulate various parts at different electrical potential. Act as a dielectric material by penetrating into solid insulator and filling spaces between the layers.

- Volatility of oil should remain low to reduce vaporization losses. Flash point of oil plays a key role during operation and service; oil temperature should be maintained well below its flash point.

There are three issues that affect chemical stability of transformer oil, viz., temperature, availability of oxygen, and presence of a metal catalyst (copper or silver). At higher temperature, decomposition of hydrocarbon molecules may take place which leads to degradation of the oil. Oxygen content in dielectric oil may lead to rise in acidity and sludge formation [10] that affects the performance of both liquid and solid insulator.

Mineral oil application in a transformer is potentially harmful to the environment during its operation, particularly in case of transformer explosion, which may lead to spillage of oil into soil and water stream. Hence, liquid insulator should accomplish minimum health and ecological necessities, viz., degradation by natural process (biodegradability), nontoxicity, reconditionable, readily disposable and nonhazardous material, reprocessability, and thermally stable.

General electric (GE) primarily adapted mineral oil as a dielectric coolant in 1892. The main reason for using mineral oil as an insulating material is its mass production worldwide and its higher flash point. In recent days, mineral oil has been widely used as a main source of insulating material for various types of equipment, especially power transformers. Poor biodegradability characteristics of mineral oil restrict its usage as a liquid insulator due to environmental concern in case of leakage during operation or by other means. Due to nonflammable properties, many transformers were insulated by a mixture of PCBs and chlorobenzenes (askarel) till 1970s [11]. Later, askarel was no more recommended as a liquid insulator in power transformer due to environmental issues and harmful to nature. In order to slow down the environmental and sustainable issues, researchers started looking for alternate liquid insulators. Very recent insulating oil is a biodegradable vegetable oil-based fluid and is popularly known as a possible alternative to replace the existing mineral oil. In 1962, for the first time, natural esters was used as an insulator in capacitor and proved a good compatibility with cellulose and other components of a transformer due to its higher dielectric constants [12].

Vegetable oils are perfect substitute to generally used mineral oil and synthetic liquids (relatively expensive). Bio-based oils have the potential to reduce environmental risk due to their completely biodegradable and nontoxic in nature. They are inexpensive than synthetic oils, and in long run, they may be even low cost than mineral oils. Furthermore, their application as a liquid insulator in power transformer contributes to defend global reserves. The drawback with vegetable oil as an insulating oil is its high pour point, higher viscosity, and more prone to oxidation. Some vegetable oils consist of high content of unsaturated fatty acids, which results in lower viscosity and improved low-temperature properties. Others have a higher percentage of saturated acids (improved oxidation stability). An optimal balance has to be chosen for the types of fatty acid contents in seed oils. There are several vegetable oils available that shows the insulating properties (see Table 6.2) to be used in transformer fluid and detailed discussion as follows:

6.3.1 *Soybean Oil*

A variety of vegetable oils available in nature, among them soybean oil, is the prominent bio-based fluid extracted from soybean seeds (*Glycine max*). Soybean oil mainly consists of unsaturated fatty acid similar to sunflower oil. The most abundantly available unsaturated fatty acids in soybean oil are polyunsaturated α-linolenic acid (7–10%) and linoleic acid (51%) and monounsaturated oleic acid (23%). It also comprises of saturated fatty acids (stearic acid, 4%) and palmitic acid (10%) [13]. Soybean oil is a major bio-source with abundant applications in the current scenario and is categorized as linolenic acid oil as it comprises of highly unsaturated linolenic acid.

Table 6.2 Typical data of different insulating fluids

Properties	Test method	Natural ester	Vegetable oil	Synthetic ester	Mineral oil
Kinematic viscosity, cSt	ISO 3104				
(1) 0 °C		84	200	350	55
(2) 40 °C		17	40	30	10
(3) 100 °C		4.6	9	5	3
Pour point, °C	ISO 3016	−30	−15	−60	−55
Flash point, °C	ISO 2719	175	290	270	140
Fire point, °C	ISO 2592	>200	330	310	180
Electric strength (kV)	IEC 60156	74	55	60	65
Tandelta at 90 °C	IEC 60247	0.04	0.02	0.02	0.0001
Relative permittivity at 90 °C	IEC 60247	2.8	3.0	3.2	2.2
Density at 20 °C (kg/m^3)	ISO 3675	890	930	970	890
Acid value, (mg of KOH/g)	IEC 62021–1	0.12	0.05	0.02	0.01
Water permeation at 20 °C (mg/kg)	–	1000	2000	1500	50
Thermal conductivity at 20 °C (W/m/K)	ASTM D2717	0.22	0.17	0.16	0.14
Specific heat at 20 °C (kJ/kg$^-$K^{-1})	ASTM D2766	2.02	1.9	1.95	1.9
coefficient of expansion (10^{-3} K^{-1})	ASTM D1903	0.75	0.70	0.74	0.75
Biodegradability after 28 days (%)	OCDE 301 B	>85	>90	–	<40

Table 6.3 Properties of soybean oil based vs. mineral oil

Properties	Mineral oil	Soybean oil
Dielectric strength (kV)	60	39
Moisture content (mg/kg)	1.5	1.9
Pour point, °C	−40	−1
Flash point, °C	154	242
Density, g/cc at 15 °C	0.895	1.462
Viscosity, mPa/s at 100 °C	1.3	29.2

In natural form, soybean oil is oxidative unstable, and if use in a transformer, it increases its viscosity. In extreme cases, if oil is left in a system, it will polymerize leading to increase in viscosity. Soya oil on chemical modification with antioxidant showed a viscosity on the order of canola oil [14]. There are efforts on chemical modification of soybean oil as a means of increasing its oxidative stability. This led to identification of one of the most stable commercially available, chemically modified soybean oils. Soybean oil is partially hydrogenated when combined with two antioxidants, citric acid (10–1000 ppm) and tertiary butyl hydroquinone (200–1000 ppm); the oil showed significantly more stable than other soybean oils [14]. Furthermore, oil is winterized in order to improve its pour point in cold temperatures. When compared with unmodified oil, the chemically modified soybean oil showed almost 50% improvement in its viscosity stability [14].

Usman et al. [7] analyzed the properties of soybean oil and compared with mineral oil, and it was observed that soy-based vegetable esters showed superior properties over mineral oil as shown in Table 6.3. Hence, concluded soy-based natural esters can be used as a liquid insulator in low-rating transformers on proper refining and structural modification. Soybean-based oil has been effectively used in transformers [15, 16], and comparison of properties is presented in Table 6.3.

6.3.2 *Palm Oil*

Palm oil is a reddish color derived from the fruit of a palm tree (See Fig. 6.6) with high β-carotene content. Palm oil has higher saturated fat content, hence, it is available in semisolid state at room temperature. Various types of palm oils are available, viz., crude palm oil, redefined bleached and deodorized (RBD) oil, etc. The key origin for palm oil is Malaysia and Indonesia. The basic major component of palm oil is triacylglycerol (TG); 95% of palm oil contains glycerol components with three fatty acids. In normal condition, palm oil is a semisolid state; hence, viscosity of palm oil is high [17]. Naturally, palm oil is categorized as a stabilized oil due to its chemical composition.

Palm kernel mainly consists of saturated fatty acid like coconut oil (Sect. 6.3.3). Previous research investigations proved that electrical, chemical, and physical characteristics of palm oil-based vegetable fluids are superior to conventional mineral oils [18]. Palm oil has a balanced configuration of unsaturated and saturated fatty acids. Hence, palm oil-based natural esters are a better substitute for mineral oil to be used as a transformer insulating oil. If the palm oil is used as an impregnation to paper insulation in transformer application, palm oil possesses superior dielectric constant and experiences lowest electric field stress.

Palm kernel oil is one of the best substitutes to mineral oils as an insulting fluid due to the presence of fatty acids with a traces of carbon. The insulation properties of palm kernel oil can be enhanced quite significantly by oxidizing unsaturated fatty acids and carbon in a stable environment. In addition to the presence of required fatty acids and carbon for better insulation, it also has other advantages which include a higher flash point, higher dielectric strength, insulation, and partial discharge properties. It is also biodegradable and nontoxic in nature, even at normal temperature, that makes it eco-friendly. Palm kernel oil is abundantly available and low cost compared to fossil fuels.

Abdelmalik [19] has provided enough evidence that palm oil can be used as an insulator in transformers. The research investigations have shown that palm kernel oil has superior properties similar to other natural esters and has all required fatty acids that can act as an effective insulator.

There is widespread research on usage of palm oil as a substitute to insulating oil in power equipment (transformers, switchgears, capacitors) [4, 20, 21], and its dielectric properties have shown a good deal with existing requirements as per national/international standards. Palm oil also has high flash and fire points and dielectric strength (BDV); however, it was also observed that kinematic viscosity of palm oil does not justify the standards. Exclusion of this drawback should be considered for realization of utilization of palm oil as an insulating fluid in transformer. Since it has been known for its ability, blending of RBDPO with soybean oil can be used as a vegetable grade liquid insulator in transformer [22, 23].

Palm oil is a good substitute for mineral oil as a liquid insulator in power equipments. In terms of viscosity, performance of most of the palm-based esters is similar to other kinds of vegetable insulating fluids [24]. Neutralization value of palm oil is comparably higher to vegetable fluids. In order to ensure the performance, additional investigations on oxidation stability of several types of palm oils are essential in present scenario. Nevertheless, flash point of palm oil is not comparable to either vegetable or synthetic esters. Hence, further investigations are required to improve the fire-safety performance of palm oil. Typical characteristics of palm oil esters and comparison with mineral oil are presented in Table 6.4.

Ghani et al. [25] studied the influence of water content on dielectric strength and characteristics of mineral and palm oil-based transformer oils by Fourier transform infrared spectrometer technique (FTIR). In their study, moisture content of oil samples varied by addition of distilled water, ranging from 1 to 5 ml. It was found that there is a decreasing trend in BDV for both mineral oil and palm oil-based ester containing 1 ml of water, with increasing water content in oil samples from 1 to 5 mL; a decrease in BDV was observed in case of mineral oil; however, the mean BDV of palm oil ester showed an increasing trend. Breakdown voltage indicate that

Table 6.4 Comparison of palm oil esters vs. mineral oil properties

Properties	Palm oil ester	Mineral oil
Density (g/cc) at 15 °C	0.86	0.88
Kinematic viscosity, cSt at 40 °C	5.06	8.13
Flash point, °C	186	152
Pour point, °C	−32.5	−45
Neutralization value (mg of KOH/g)	0.005	<0.01
Dielectric constant	2.95	2.2
Volume resistivity, Ohm cm^{-1}	7.1	7.6
Breakdown voltage (kV)	81	70–75

the effect of moisture content is more pronounced for mineral oil samples compared to palm oil-based ones.

6.3.3 *Coconut Oil*

Coconut oil, a colorless to pale yellow liquid, is an edible vegetable oil processed from kernel of ripened coconuts gathered from coconut palm. Coconut oil is subjugated due to the presence of nearly 90% saturated fats and 10% of unsaturated fatty acids [26, 27]. In Pacific Islands, studies are being carried on the possible usage of coconut oil as a fuel for electricity generation. Due to the cost of labor and supply constraints to date, it seems that it is not useful as a fuel source [28]. The price of the coconut is relatively less compare to synthetic oil, and further, it is also eco-friendly. These would be the reason to investigate coconut oil as a substitute for synthetic oil. Coconut oil has also been tested for its use as a lubricant [29] and as a dielectric fluid in transformers [30].

Introductory studies evidenced that coconut oil can be a good alternative for transformer liquid insulation [31–35] due to its acceptable breakdown voltage levels. However, differences are noted in some of the properties, viz., interfacial tension reduces, whereas neutralization value increases and dielectric dissipation factor increases with aging. Limited studies on various features and categories of coconut oil including physical, chemical, and electrical properties were previously carried out [26–30] The literature shown that coconut oil could satisfactorily confirm with standards its performance and is equivalent to different types of natural esters as per standards [34–37]

Lucas et al. [37] studied that the performance of coconut oil filled 5 kVA (single phase) and 160 kVA (three phase) transformer and proved that on decontamination and de-miniaturization, it can be used adequately in tropical climates in sealed power transformers. Existing literature proved that coconut oil may be suitable in low-temperature climates though it freezes at around 23 °C.

Table 6.5 Physical properties of castor oil [40]

Properties	Values
Flash point, ⁰C	229
Density at 25 °C, g/cc	0.96
Boiling point, ⁰C	313
Melting point, ⁰C	−10
Solubility in water	Not soluble

6.3.4 Castor Oil

A typical vegetable oil for consideration as an alternative ester-based transformer insulation oil is castor oil; it is a yellowish-brown fluid with no odor or taste and it is derived from *Ricinus Communis* plant seeds (*Ricinus oil*). It boils at around 313 °C and density is 0.961 g/cc. It is a triglyceride in which approximately between 87% and 90% of fatty acid chains are ricinoleic acids. It also contains other components, viz., oleic and linoleic acids. Castor oil is one of the rare naturally occurring glycerides and is being a pure compound since fatty acid portion is about nine out of ten portions of *ricinoleic* [38].

The suitability of castor oil as alternative transformer insulation oil was investigated by Olawuni et al. [39] and reported that it has similar physical properties of mineral transformer insulation oils and is suitable for low-voltage transformer (11/0.415 kV) on removing impurities and chemical modification. Typical physical properties of castor oil is presented in Table 6.5.

6.3.5 Sunflower Oil

Vegetable oils based on high oleic acid content are chosen as optimal due to its relative oxidative stability. It is a mixture of poly- and monounsaturated oleic and linoleic acid group of oils extracted from *Helianthus annuus* seeds. Attempts have also been made on the suitability of sunflower oil as dielectric fluid is 100% ecological balanced, but the drawback is its cost.

6.4 Natural Ester Oil as a Liquid Insulator: A Historic Evaluation

The drawback with vegetable oil-based coolants is their high pour point and lower aging characteristics (oxidation stability) compared to conventional mineral oils [41]. Except for special claims, renewed interest in ester fluids did not occur until the 1970s (notorious issue of the PCB) united with oil crises in coming years, lawful the prerequisites of renewable transformer oil.

The mid-1980s experienced the necessity for power equipment and appliances using vegetable oils as an insulating media. Due to embraced dimensions, transformers had forced circulation flow to isolate heat exchangers. Therefore admirable lubrication, inferior pour point, and predominant fire points were important liquid properties for their end usage. High cost compared to other liquid dielectrics is a main drawback for accepting synthetic esters for the intended purpose [42]. An extensive research was initiated in the 1990s due to environmental concern and accountability risks involved in non-edible oils which led to reentering the natural esters. They showed excellent insulating and fire-safety characteristics of synthetic polyol esters, and they are categorized as fit for human consumption. In addition, they are bio-decomposable due to presence of an organic composition, and they are more economically viable than synthetic esters [43, 44].

Due to their "green" credentials, there has been a rush for past one and half decade in the use of natural ester liquid insulators. Currently existing commercial natural esters based on oleic oils are "BIOTEMP"® and "Envirotemp FR3"® [45, 46]. In 2007, Siemens has filled 24,000 V, 130 MVA transformers with model 7131 for European utility and ABB has filled 130 kV, 25 MVA transformers with BIOTEMP in Brazil and Egypt [47]. Alstom has filled several transformers (130 kV, 90 MVA) and reactors (up to 220 kV) with FR3.

Power transformers have grown for past few decades in terms of various characteristics, viz., ecological balance, technology, performance, and power ratings. Today, power industry is looking for shifting from harmful era and unfriendly environment petroleum-based mineral oils that are going to depict in the near future to natural esters, a renewable, environmentally friendly, and nonhazardous. In the past few decades, there has been a retrieval of usage of natural ester fluids for "green" and save credentials. One prominent manufacturer highlights the safety of synthetic esters, as no fire cases have been reported, while another US manufacturer indicates that over 100 utilities are using natural esters in distribution transformers. Field studies performed using FR3 in several distribution transformers since 1996 have concluded successful operation [48].

Mineral oil is a most commonly used liquid insulator for many years in transformer applications, and a wide-range database of knowledge has been accrued over the years. Increased environmental, safety, performance, and economic have led to consideration of other oils to use as a transformer insulating fluids. Since discovery of oil-filled transformers in the late 1880s, vegetable esters have been used as dielectric liquids. The earliest natural esters were found to be incompatible with free-breathing equipment, because of their oxidation characteristics, and were gradually replaced by mineral oils [49]. Environmental requirements have caused electrical power sector to look for viable alternatives to mineral oil. To be acceptable, any environmental friendly alternative must be benign, cost-effective, and offers a high electrical performance in long run. Recently there is a restoration of usage of vegetable esters due to "green" credentials [49, 50].

Mineral oil is non-biodegradable and contaminates soil and water. By mid-twenty-first century, petroleum products are ultimately going to run out due to severe scarcity. Hence, researchers worldwide are earnestly thinking of alternate insulating fluids from natural resources. Keeping in view of the keen interest towards biodegradable dielectric fluids, research efforts were started to develop fully biodegradable liquid insulators in the 1990s. Abundant availability of vegetable oil in a natural form was considered as the most likely candidate for biodegradable dielectric fluids and is fairly good insulator with superior thermal properties.

The scientists documented that vegetable oils to be used as a liquid insulator need further development. Since fluid in a sealed transformer should remain for 30–40 years, long-term stability of transformer insulating oil is of critical importance. Natural oils intrinsically have components that degrade in a fairly short period. The rate of unsaturation from mono- to tri-unsaturation is an indication of thermal instability of the liquid.

In one significant project, an Asian utility serving a significant customer base has recently specified smaller, more compact transformers utilizing natural ester technology, anticipating saving up to 20% of their annual transformer budget. Natural ester-filled power transformers would not be complete without reviewing the largest unit installed to date. In early 2014, Envirotemp™ FR3™ fluid, a soy-based natural ester insulating liquid, was filled in 420 kV extra high-voltage transformer.

6.5 Natural Esters vs. Mineral Oil

Natural ester (vegetable oil-based) insulating fluids have superior properties over conventional petroleum-based fluids, viz., thermal performance, fire safety, environmental benign, etc. These liquids are self-extinguishers (non-propagating) and can be used in various industrial applications, namely, safety device to space separation, fire barriers, and extinguish systems which diminish a fire but do not prevent it. Interaction between vegetable esters and cellulose insulation gives longer insulation life and allows higher or extended overloads without abnormal loss of insulation life. Vegetable oils can have lower or no toxicity and have good environmental characteristics superior to those of mineral oil and are commonly used as food products. The details are given in Table 6.6.

Due to its low oxidation stability, mineral oil can produce toxic substances, discarding and clearance after self life and on spillage are complex and lead to serious threat to environment. Due to the above reasons, its usage in many countries is highly doubtful. Vegetable oils undergo complete and rapid degradation in presence of microorganisms and release very low toxics [51]. This is primarily due to the fact that natural esters do not consist of volatile organics, halogens, polynuclear aromatics, and other complexes that are existing in petroleum-based oils [52]. Various researchers reported that natural ester fluids undergo 70–100% biodegradation in 28 days [53]. In both aerobic and anaerobic conditions, vegetable oils have a better biodegradation capacity.

Most of the natural esters suffer from not being a stable for oxidation as other type of liquid insulators. Although natural esters produce from a wide variety of

Table 6.6 Typical properties of mineral transformer oil vs. vegetable oils [19]

Properties	Conventional mineral transformer oil	Vegetable oil with saturated fatty acid 80%	Vegetable oil with unsaturated fatty acid >80%
Viscosity (cSt) at 40 °C	13	29	32.6
Density (g/cc) at 20 °C	0.895	0.917	0.886
Pour point (°C)	−40	20	−22
Flash point (°C)	154	225	260
Oxidation onset temperature (°C)	207	282	192
Conductivity (S/m) at 20 °C	10^{-13}	10^{-11}	10^{-10}
Breakdown voltage (kV)	45	60	56

plant-based oils, natural esters for insulating applications are most commonly developed from soy and sunflower oil. This may be due to abundantly available, low cost, and superior performance. Breakdown voltage, thermal conduction, and oxidation stability are the minimum set of tests to be checked to analyze any oil to be used as a liquid insulator in power equipment. Under AC voltage circumstances, natural ester behavior is equivalent to mineral oil, and breakdown voltage of solid insulation infused with natural esters is same as that of infused with mineral oil.

Vegetable esters have high thermal and chemical stability with a fire point greater than 360 °C and classified as "K" class. Vegetable oil-filled transformers have practical installation and regulatory advantages in many sites worldwide [54–56]. A sequence of trials established that thermal degradation of paper insulation in vegetable fluids decreases by a factor of 5 to 8 times compared to mineral oil [57–59]. Vegetable oils show some stability toward oxidation, and the resultant biproducts of mineral oil can undergo reaction to form sludge or sediment particles. Nevertheless, oxidization of natural esters differs from mineral oils. The oxidation biproducts of natural esters do not form any precipitants. However, oil ultimately leads to increase in viscosity due to polymerization resulting in the formation of a self-sealing structure for cable applications [60].

Various scientists have documented that the universal conclusion about aging characteristics of oil is by change in color [61], oxidation [62], and raise in neutralization value, moisture content [63] and increased dielectric loss [64]. Hosier et al. [65] concluded that dielectric dissipation factor of natural esters after certain period of aging is not worse than mineral oil; and hence they could be substituted by mineral oil for high-voltage applications. The advantages of ester fluid over mineral oil can be recognized from its performance point of view [66] as shown below:

- Higher flash point, low vapor pressure, and volatiles.
- High solvency and lubricity.

- Higher polarity of ester leads to high affluence at inner parts of the fluid.
- Hygroscopic in nature and attach easily due to affinity toward polar ester bonds.
- Thermally more stable due to higher bond energy of ester linkages [67].
- Hydrolytic stability.

Natural ester fluid needs some additives such as oxidation inhibitors or metal deactivators, to improve the performance and oxidation stability. Natural esters ensure that combination between natural ester fluid and additive material is in the range of minimum environmental and health safety target.

6.6 Research Scenario

Growing demand worldwide for electrical energy, environmental and safety concerns cannot be ignored. Transformers being vital elements in power transmission network also have to follow this trend. Mineral oil has been used as a cooling agent and as a dielectric fluid for over 10 decades [68, 69] in power transformers to serve as heat-dissipating agent and insulating component [68]. Nowadays, substantial efforts were put in a search for more eco-friendly alternatives to petroleum-based insulating oil. Natural esters or vegetable oils, processed from plant-based materials, are gaining attention in recent years as an alternative to mineral oils. Vegetable fluid-filled power equipment has increased attention in distribution transformers in various countries, viz., Asia, the USA, Brazil, and other European countries; they are currently under research and development stage to use in higher-rating power transformers. Vegetable oils such as sunflower and canola have been tested and found suitable as an alternative to mineral insulators to be used in transformers [12].

Vegetable oils are believed to be the best alternative to mineral oil in transformer industry. Natural ester oils are organically produced from animal and plant origin (seeds and flowers). Several industries and researchers have keen interest on the suitability of vegetable oils an insulating medium in power transformers [68]. Natural esters with higher concentrations of unsaturated fatty acid make them unstable and susceptible to oxidation [70]. The fatty acid hydrocarbon chains and their degree of unsaturation influence the dielectric and physiochemical properties of natural esters. Hydrolysis reaction and difference in chemical structure in vegetable oil leads to formation of acids (no such reaction in the case of mineral oils); hence, acidity of vegetable oils has higher values than mineral oils [71]. Natural esters mainly consist of high molecular weight acids (stearic and oleic acids), whereas mineral oil contains low molecular weight compounds (acetic, formic, and clavulanic acids) [72–74].

On the interest of utility, research efforts started in the mid-1990s to improve fully biodegradable liquid insulators. These fluids are considered as a biodegradable and good insulating fluid for transformer applications and arose as an increasingly common mineral oil substitute [75]. They compensated all the foremost jeopardies such as high flammability and environmental impact associated with present mineral oil. They are made of renewable biological sources such as vegetables. The usage of natural ester fluids is environmentally benign due to its biodegradability and nontoxicity and possesses low emission to the environment.

Since about one and half decade, vegetable fluids have been industrialized in the USA. Standards within IEEE and IEC are available [76, 77]. Presently, there is a prodigious deal of interest in this area with different designs (a type of base oil, addition of additives, biodegradability, and fire-safety issues). Studies on long-standing performance of natural ester fluids in transformers had remarkable results; most striking was that it greatly extended the life of cellulose-based insulating paper in transformers. Over time, that insulating paper normally becomes brittle, and cellulose begins to break down, leading to equipment malfunction. Tests showed that natural ester fluids extended paper life by 4 to 8 times when it is aged in fluids. Thus, the use of natural ester fluids leads to a lower cost of the transformers over a period of cycle. In addition, risk of transformer fire is significantly reduced because of high fire point of natural ester fluids. Finally, natural ester fluids are completely biodegradable, making them an environmentally sensitive application and potentially saving electrical utilities millions of dollars towards cleanup costs. Shah et al. [78] reported that due to the occurrence of nonpolar triglycerides, dielectric constant of cottonseed and corn fluids is greater than mineral oil.

6.7 Conclusion

The use of natural esters as an insulating fluid in power transformers plays a pivotal role in supporting power industry by reducing pollution raised due to usage of mineral oils. The advancement of natural ester fluids accomplishes present necessities for eco-friendly insulating fluids in power transformer. In addition to environmental benefits, petroleum products are soon going to come to an end in the near future, and there could be serious scarcity by the mid-twenty-first century. Lack of environmental friendliness with current mineral insulating oil has prompted scientists to study further. Discovery of vegetable esters and its dielectric properties is now quite evident that future transformers are more likely to use natural ester oils in insulation applications. Fortuitously, foundation has already laid down to develop suitable dielectric fluids. Other possible application for these liquids includes cables, circuit breakers, capacitors, and tap changers and needs advanced studies and test methods.

References

1. Eberhardt R, Muhr HM, Lick W, Baumann F, Pukel G (2008) Comparison of alternative insulating fluids, IEEE-Annual Report Conference on Electrical Insulation Dielectric Phenomena, October 26–29, 591–593
2. Mc Shane P (2001) Relative properties of the new combustion-resistant vegetable-oil based dielectric coolants for distribution and power Transformers. IEEE Trans Ind Appl 37(4):1132–1139
3. Erhan SZ, Adhvaryu A (2002) Vegetable-based base stocks. In: Erhan SZ, Perez JM (eds) Bio-based industrial fluids and lubricants. AOCS Press, Champaign, pp 1–19
4. Suwarno F, Stitinjak, Suhardi I, Imsak L (2003) Study on characteristic of palm oil and it's derivative as liquid insulating materials department of electrical engineering, Bandung Institute of Technology, Bandung, Indonesia. Proceed of the 7th International Conference on Properties and Applications of Dielectric Materials, June 1–5, Nagoya
5. Gunstone FD (2004) The chemistry of oils and fats: sources, composition, properties, and uses. Blackwell Publishing Ltd, Oxford
6. Yves B, Patrick L (2013) 22nd International conference on electricity distribution Stockholm, 10–13 June, Paper 0382
7. Usman, MA, Olanipekun, OO. Henshaw, UT. J Emerg Trends Eng Appl Sci (JETEAS) 3(1):33–37.
8. Haribabu S, Hockessin Qiu, W (2014) Blended oil compositions useful as dielectric fluid compositions and methods of preparing same, Pat. No.: US 2014/0131639, 15
9. Willis HL, Welch GV, Schrieber RR (2001) Aging power delivery infrastructures, vol 12. Marcel Dekker, Inc., New York
10. Crine JP (1988) Newly developed analytical techniques for characterization of insulating oils. In: Erdman HG (ed) Electrical insulating oils. ASTM, Philadelphia
11. Fofana IV, Wasserberg H, Borsi GE (2001) Retrofilling conditions of high-voltage Transformers. IEEE Electr Insul Mag 17(2):17–30
12. Oommen TV (2002) Vegetable Oils for Liquid-Filled Transformers. IEEE Electr Insul Mag 18(1):6–11
13. Dušica IS, Jovanka LD, Slavica SA (2010) Fatty acid composition of various soybean products. J Inst Food Technol Novi Sad 37(2):65–70
14. Cannon GS, Honary LAT, (2000) Soybean based transformer oil and transmission line fluid US 6159913 A
15. Bremmer BJ, Larry P (2008) Biobased lubricants market study
16. Ursula B, Jürgen MO (2007) Faculty of mathematics and natural sciences, Carl Von Ossietzky University, Oldenburg. Application of Vegetable Oil Based-Fluids as Transformer Oils
17. Reza Z, Azmi Z, Hossein AA, Amir RS, Mohd A (2010) Fabrication of silver nanoparticles dispersed in palm oil using a laser. Ablation. Mahdi Int J Mol Sci 11:4764–4770
18. Rajab A, Sulaeman A, Sudirham S, Suwarno ITB (2011) J Eng Sci 43(3):191–208
19. Abdelmalik AA (2014) Sustain Mater Technol 1–2:42–51
20. Abdullah UU, Bashi SM, Yunus R, Mohibullah, Nurdin A (2004) The potentials of palm oil as a dielectric fluid. IEEE, National Power & Energy Conference
21. Aditama S. Dielectric properties of palm oils as liquid insulating materials: effects of fat content. IEEE Electrical Insulating Materials, 2005. (ISEIM 2005). Proceedings of 2005 International Symposium
22. Patent Genius US Patent 6613250 – Vegetable oil based dielectric fluid and methods of using same, 2006
23. Cargill Incorporated Cooper Power Systems and Cargill Industrial Oils Form Alliance to Produce Revolutionary Insulating Fluid for Electric Utilities, 2006
24. Norhafiz Azis, Jasronita Jasni, Mohd Zainal Abidin Ab Kadir, Mohd Nazim Mohtar, Suitability of palm based oil as dielectric insulating fluid in Transformers J Electr Eng Technol Vol. 8, 742 2013

25. Ghani SA, Muhamad NA, Chairul IS, Jamri N (2016) J Eng Appl Sci 11(8)
26. Matharage BSHMSY, Fernando MARM, Tuncer E, Bandara MAAP, Kalpage CS (2013) Coconut oil as Transformer liquid insulation-ageing and simulated thermal and electrical faults. IEEE Trans Dielectr Electr Insul 3:887–898
27. Coconut oil transport information service, German Insurance Association, Berlin. 2015
28. Economics of rural renewable energy technologies. Secretariat of the Pacific Community – Applied Geoscience and Technology Division. 2010–06-14. Retrieved 2011–08-09
29. Romares-Sevilla, J (2008) Davao-based firm sees an expansion of bio-tech oil market. Sun. Star Superbalita Davao. Archived from the original on 2008–01-21. Retrieved 14 July 2008
30. Abeysundara DC, Weerakoon C, Lucas JR, Gunatunga KAI, Obadagee KC (2001) Coconut oil as an alternative to transformer Oil. ERU Symposium
31. Kurera NASA, Perera KAR, Ranawana S, Ekanayake CMB, Fernando MARM, Analysis of insulation characteristics of coconut oil as an alternative to the liquid insulation of power transformers, IEEE ICIIS-2008, Paper No. 178, Kharagpur, Dec. 2008
32. Hinduja D, Gayathri CSK, Fernando MARM (2011) Laboratory investigation of treated coconut oil as transformer liquid insulation. IEEE ICIIS2011, Kandy, pp 108–113
33. Lucas JR, Abeysundara DC, Weerakoon C, Perera KBMI, Obadage KC, Gunatunge KAI (2001) Coconut oil insulated distribution transformer. In: IEEE 28th annual conference Sri Lanka, pp 1–5
34. Yusnida MY, Kamarol M, Ahmad MA (2012) Breakdown voltage characteristic of MO and RBDPO mixture for power transformer insulation. IEEE Colloquium Hum Sci Eng (CHUSER) 3–4:300–303
35. Ranawana S, Ekanayaka CMB, Kurera NASS, Fernando MARM, Perera KAR (2008) Analysis of insulation characteristics of coconut oil as an alternative to the liquid insulation of power transformers. IEEE 3rd International Conference Industrial and Information Systems (ICIIS 2008) pp 1–5
36. Bertrand Y, Lauzevis P (2013) Development of a low viscosity insulating liquid based on natural esters for distribution transformers. 22nd International Conference on Electricity Distribution (CIRED) 10–13:1–4
37. Lucas J. R, Abeysundara DC, Weerakoon C, Perera KBMI, Obadage KC, Gunatunga K.A.I, Paper 2001/02 presented at the 8th Annual conference of the IEE Sri Lanka on 22 Sept 2001
38. Akan UG, Jimoh A, Mohammed AD (2006) Extraction, characterization and modification of Castor seed oil. Leonardo J Sci 5(8):43–52
39. Olawuni A, Adejumobi IA (2014) Int J Sci Eng Res (IJSER) 2:2347–3878
40. Dairo OU, Alamu OJ, Olayanju TMA, Ajisegiri ESA. and Awonorin, OO. (2010) Optimization of in situ biodiesel production from raw castor oil-bean seed using response surface methodology. HumboldtKolleg International Conference (Humboldt-Kolleg Ogbomoso 2010) held at the Ladoke Akintola University of Technology, Ogbomoso (Jan.11–14, 2010) Book of Abstract p 22
41. Wilson AC (1980) Insulating liquids: their uses, manufacture and properties. Institution of Electrical Engineers, London
42. Oommen TV (2000) A new vegetable oil based transformer fluid: development and verification. Proc. CEIDP, Victoria, pp 308–312
43. Lucas EW, Rhee KC (1992) Animal and vegetable fats, oils and waxes. In: Kent JA (ed) Riegel's handbook of industrial chemistry, 9th edn. Van Nostrand-Reinhold, New York. Ch. 8
44. Shinke M, Kenji M, Toshiharu T, Yasuo T, Yoshitake N, Shimizu R, Kosaka M and Wada M (2003) Fundamental studies on the development of environmental friendly vegetable oil filled transformer, Magazine of insulation and dielectrics IEE Japan
45. Vishal S, Saurabh P, Vikas and Prashant, (2011) Transformer's history and its insulating oil. Proceeding of 5th National Conf. (INDIACom), ISSN 0973–7529, ISBN 978–93–80544-00-7
46. Mc Shane CP (2002) Vegetable-oil-based dielectric coolants. IEEE Ind Appl Mag 8(3):34–41
47. Boss P (2007) New insulating fluids for transformers based on biodegradable high oleic vegetable oil. Matpost Conf, Lyon

48. Rapp K, Stenborg P (2004) Cooper power systems field analysis of envirotemp FR3 fluid in sealed versus free-breathing transformers, CP0414, Cooper Power Systems
49. Mc Shane CP (2001) IEEE Trans Ind Appl 37(4):1132–1139. No. 0093-9994/01
50. The Environmental technology verification program, U.S. Environmental Protection Agency, Washington, DC, VS-R-02-02 (June 2002)
51. Honary LAT (2004) Biodegradable/Biobased lubricants and greases, Machinery Lubrication Magazine Issue No. 200109, Noria Corporation
52. Howell S (2007) Promising industrial applications for soybean oil in the US
53. Broekhuizen P, Theodor D, Le Blanch K, Ullmer S (2003) Lubrication in inland and coastal water activities. A. A. Balkema Publishers, Tokyo
54. Transformer fire protection, facilities instructions, standards, and techniques, Vol. 3–32, Hydroelectric Research and Technical Services Group, Bureau of Reclamation, U.S. Dept. of Interior (Denver, CO USA, Jan., 2005)
55. 5–4 Transformers, Section 2.3.1.1.1 Separation Distance, Property Loss Prevention Data Sheets, Factory Mutual Global, May (2003), Norwood
56. Classification of insulating liquids according to fire-point and net calorific value (2005) CEI/IEC 61100: 1992–05, International Electrotechnical Commission, Geneva p 11
57. Mc Shane CP, Rapp KJ, Corkran JL, Gauger GA, Luksich J (2001) Aging of paper insulation in natural Ester dielectric fluid. EEE/PES Transmission & Distribution Conference, Atlanta
58. Mc Shane CP, Rapp KJ, Corkran JL, Luksich J (2003) Aging of Cotton/Kraft Blend Insulation Paper in Natural Ester Dielectric Fluid Tech Con 2003 Asia Pacific, Sidney, pp 7–9
59. Mc Shane CP, Corkran JL, Rapp KJ, Luksich J (2003) Aging of paper insulation retro filled with natural ester dielectric fluid. In IEEE/DEIS Conference on Electrical Insulation and Dielectric Phenomena, Albuquerque
60. Standard Specification for Natural (Vegetable Oil) Ester Fluids Used in Electrical Apparatus, ASTM D6871–03. ASTM International, West Conshohocken
61. Ferguson R, Lobeiras A, Sabau J (2002) Suspended particle in the liquid insulation of aging power transformer. IEEE Electr Insul Mag 18(4):17
62. Blue R, Uttamchandani D, Farish O (1998) Infrared detection of transformer insulation degradation due to accelerated thermal aging. IEEE Trans Diel Electr Insul 5:165–168
63. Rouse TO (1998) Mineral insulating oil in transformers. IEEE Electr Insul Mag 14:6–16
64. Kanno M, Oota N, Suzuki T, Ishii T (2001) Changes in ECT and dielectric dissipation factor of insulating oils due to aging in oxygen. IEEE Trans Diel Electr Insul 8:1048–1053
65. Hosier IL, Vaughn AS, Swingler SG (2008) Studies on the aging behavior of various synthetic and natural insulation oils. In IEEE International Conference on Dielectric liquids 29 June – 4 July
66. Randles SJ (2006) Esters. In: Rudnick LR (ed) Synthetics, mineral oils, and Biobased lubricants: chemistry and technology. CRC Press, Taylor & Francis Group, Florida
67. Bhushan B (2002) Introduction to tribology. Wiley, New York
68. Adejumobi IA, Oyagbirin SG (2007) Insulation deterioration and its effects on power systems, journal engineering and. Appl Sci 2(5):870–873
69. Yang X (2014) Oxidation stability assessment of a vegetable transformer oil under thermal aging. IEEE Trans Dielectr Electr Insul 21(2):683–692
70. Abdel M, Fothergill JC, Dodd SJ, Abbott AP, Harris R (2011) Effect of side chains on the dielectric properties of alkyl esters derived from palm kernel oil. In Proceedings of the IEEE international conference on dielectric liquids (ICDL), pp 1–4
71. Liao R, Hao J, Chen G, Ma Z, Yang L (2011) A comparative study of physicochemical, dielectric and thermal properties of pressboard insulation impregnated with natural ester and mineral oil. IEEE Trans Dielectr Electr Insul 18(5):1626–1637
72. Azis, N and Wang, ZD (2011) Acid generation study of the natural ester. In Proceedings of the XVII international symposium on high voltage engineering, Hannover; August 22–26
73. Lundgaard LE, Hansen W, Ingenbrigtsen S (2008) Ageing of mineral oil impregnated cellulose by acid catalysis. IEEE Trans Dielectr Electr Insul 15(2):540–546

74. Gómez NA et al (2014) Dissolved gas analysis (DGA) of natural ester insulating fluids with different chemical compositions. IEEE Trans Dielectr Electr Insul 2014:21(3)
75. Oommen, TV, (1999) Biodegradable transformer fluid from high oleic vegetable oils. Doble conference paper; April.
76. IEC 62770 ed1.0 (2013) Fluids for electro technical applications - Unused natural esters for transformers and similar electrical equipment
77. IEEE Guide C57.147 (2008) - IEEE Guide for Acceptance and Maintenance of Natural Ester Fluids in Transformers
78. Shah ZH, Tahir QA (2011) Dielectric properties of vegetable oils. J Sci Res 3(3):481–492

Chapter 7
Properties of Vegetable Fluids: A Green Insulator for Power Sector

Abstract This chapter deals with the properties (physical, thermal, chemical, and electrical) of natural esters to be used as insulating fluids. This chapter also describes the effect of various additives (antioxidant additive, pour point depressant, etc.) on the properties of vegetable oil-based natural esters. It also gives detailed discussion on current research scenario of vegetable oil-based ester fluids for power sector. It also includes challenges and technical difficulties of vegetable ester fluids as a liquid insulator; it also covers comparison between vegetable oil fluids and mineral oils as liquid insulators for transformer application.

Keywords Natural ester • Biodegradable • Electrical properties • Renewable source • Gassing tendency

7.1 Introduction

For more than half a century, most of the power transformers have been used with mineral oils. Due to its availability, enhanced properties, good compatibility with a cellulose paper insulation, and low cost, substantial use of petroleum-based mineral oils has been justified. Recently, environmental concerns become more imperative, usage of oil having higher fire point and biocompatibility becoming more attractive. Thus, currently available vegetable oil (a renewable source)-based natural ester fluids have provided a substitute to mineral oils for use in transformers as an insulating fluid and cooling agent. It is an interesting challenge converting from mineral oil to vegetable oil, due to distinguished properties that affect both design and manufacturing processes. To proceed further, it is essential to understand the variations that predominantly occur in terms of insulation, thermal properties, processing, diagnostic, and cost issues.

Nowadays, power equipment (transformers, switch gears, capacitors, etc.) are filling with various liquid insulators, viz., silicon, mineral oils, and synthetic and vegetable esters, depending on the usage of each type on justifying its end use. Due to the increasing call for use of eco-products in power sector, various power industries are working on the development of usage of vegetable esters in power

D. Gnanasekaran, V.P. Chavidi, *Vegetable Oil based Bio-lubricants and Transformer Fluids*, Materials Forming, Machining and Tribology,
DOI 10.1007/978-981-10-4870-8_7

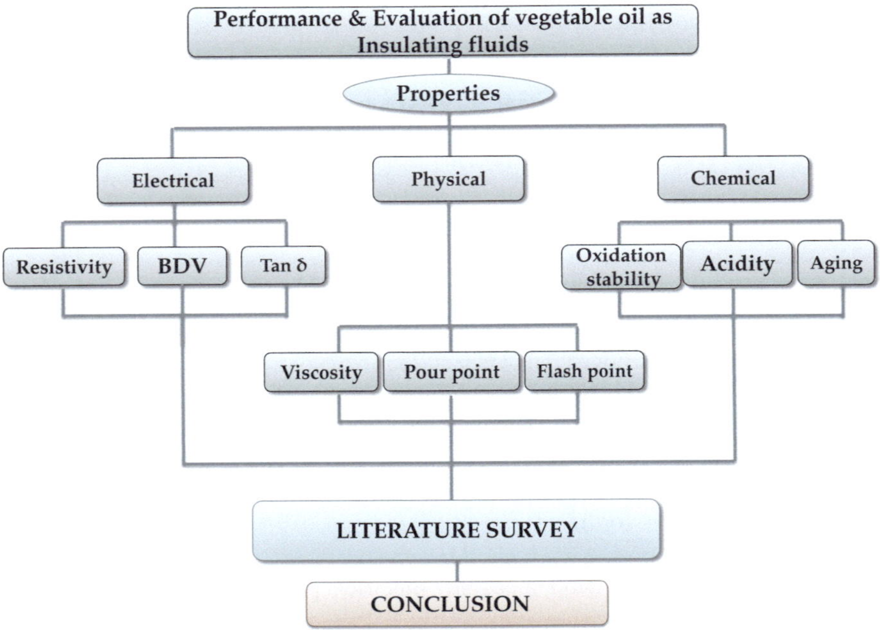

Fig. 7.1 Essence of the chapter

transformers. The content of this chapter is briefly presented in Fig. 7.1, and the detailed discussion is as follows.

7.2 Properties of Natural Ester Fluids

Vegetable ester fluids are processed from various plant-based seed oils. Physical and chemical properties of ester liquids depend on sources that are available, chemical structure, and saturation/unsaturation ratio of fatty acids. Natural esters exhibit better oxidation stability if composition is having higher percentage of monounsaturated oleic acid (see Sect. 6.2 and Table 6.1).

Vegetable esters are used in power equipment because of their insulating properties that restrict due to their failure for high-voltage application. The impurities (humidity, gasses, and particulates) present in the oil highly influence electric strength, which leads to the tendency of comparing them with conventional mineral oil by monitoring impurity content. Water solubility of natural esters and mineral oils makes great difference among them in terms of moisture-absorbing capacity; natural ester fluids have a higher water-absorbing capacity (20–30 times more) compared to mineral oil that influences effect of humidity on breakdown voltage [1] (see Table 6.2). Important characteristics of various insulating oils are given in Table 7.1. Hence, ester oils show superior properties than petroleum-based mineral oils.

Table 7.1 Characteristic properties of mineral oil and natural ester fluids [2]

Properties	Mineral oil	Natural esters
Source	Purified from crude petroleum	Extracted from crop plants
Main component	Complex mixture of hydrocarbons	Plant-based natural esters of triglycerides
Chemical structure	*Paraffin/isoparaffin* 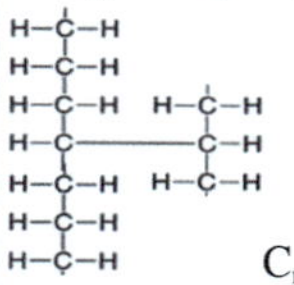 C_nH_{2n+2} *Naphthenic* 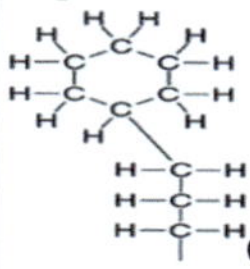 C_nH_{2n} *Aromatic* 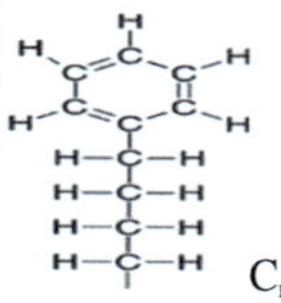 C_nH_{2n-6}	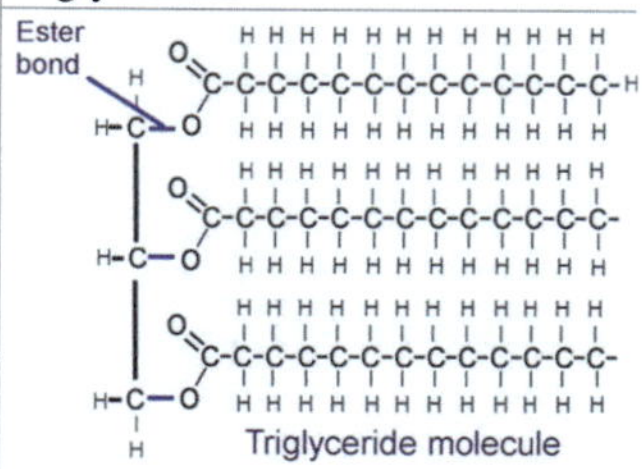
Biodegradability	Slow biodegradation	Readily biodegradable
Oxidation stability	Good stability	Oxidation susceptible
Water saturation at ambient temperature (ppm)	55	1100
Flash point °C	160–170	>300
Fire point, °C	170–180	>350
Breakdown voltage, (kV)	74	55
Fire safety classification	O	K

Note: O-flash point <300 °C, K-flash point >300 °C as per IEC 61100 standard
Vegetable oils classed as low-flammability fluids as per IEC standards

In terms of safety, environmental issues, and thermal properties, natural ester fluids have superior properties compared to mineral oil. These are characterized as high fire point fluids with self-extinguishing (non-propagating) properties and suitable to use in various industrial applications. Thermal properties and interaction with solid insulation (kraft paper) give long insulation life and allow higher or extended overloads, over an extended period of time without unusual loss of dielectric properties.

Vegetable ester fluids are considered as the best alternative to mineral oils where fire safety, environmental hazards, or superior insulation properties are required. Fire safety is a key concern of today's dielectric material users, especially their use in subway tunnels or ships; this also applies where they will be used in populated areas such as offices, shops, and workplaces. There are many instances that explosion

Table 7.2 Electrical properties of natural ester fluids vs mineral oil [4]

Dielectric properties	Mineral oil	Natural ester fluid
Electric strength (BDV), kV	54.9	56.7
Dielectric dissipation factor	0.081	0.45
Specific resistance, 10^{12}ohm cm^{-1}@ 80 °C	220	3.0
Gassing tendency, μl/min	−5	−79

of transformers result in external fires which are difficult to extinguish and spread to surroundings as the oil spills on the ground. Typical properties of mineral oil and vegetable-based natural esters are presented in Table 7.1.

7.2.1 *Electrical Properties*

Insulating properties of oil vary depending on various factors such as water content, temperature, applied voltage, frequency level, polarity, electrode geometry, particles, etc. The use of vegetable-based esters as a substitute to mineral oil has many advantages. The permittivity of natural ester fluid affects the permittivity of solid insulation impregnated with ester fluid that is higher than that of mineral oil. Due to the closer permittivity values of vegetable fluids and impregnated paper insulation, a higher stress is experienced on paper insulation in case of vegetable ester fluids than mineral oils [3]. To select suitable electrical insulation liquids, it is necessary to know the dielectric properties: electric strength (BDV), dielectric constant (DC), dielectric dissipation factor (DDF), specific resistance (SR), gassing tendency (GT), etc. Comparison of electrical properties of different types of insulating oils is given in Table 7.2.

7.2.1.1 Electric Strength (BDV)

AC breakdown voltage is the most important parameter that an insulating fluid must meet. The AC breakdown voltage is highly sensitive to excess of moisture, particulates, and air or gas bubbles (impurities) present in a transformer fluid; hence, measured AC breakdown voltage represents quality of oil rather than its characteristics, and highly associated with oil chemistry, it forms good properties when the impurity content is strictly controlled. The major purpose of filling transformer with oil is to provide electrical insulation; insulation ability is controlled by complex oil/paper composite system. Oil permeates into solid insulation and thus pushes away air (a lower dielectric strength than oil). Electric strength is one of the most important properties used to determine the efficacy of liquid dielectrics.

Darwin et al. [5] studied the suitability of natural esters as an insulating fluid for transformers and compared the properties with mineral and synthetic esters. In previous studies, it was observed that there is a large variation in water-soluble level between natural ester oils and mineral oils. Superior moisture solubility reduces the effect of humidity on strength of insulation but also keeps solid insulation dry leading to enhance shelf life of a transformer (life of solid insulator is precise by condition of solid insulation). On the other hand, vegetable esters induce water quickly, and extra care has to be taken during the usage of transformers due to its higher moisture solubility [5].

Natural esters need discrete operating conditions involving cleaner and sealed apparatus. In these circumstances, natural esters showed the same or better insulating characteristics than mineral oil [6]. In a recent research work carried out by Martin et al. [7], dielectric strength of cellulose paper impregnated in ester fluid showed almost equal to that of mineral oil.

7.2.1.2 Dielectric Dissipation Factor (DDF)

Dielectric dissipation factor in other words tan delta is a tangent of dielectric loss angle during the applied voltage (V) of frequency (f) across an insulating system comprising of capacitance (C). This is an angle in which phase variance between applied voltage and substantial current deviates from $\pi/2\o$, at a point where capacitor's dielectric includes only insulating material.

Loss angle is a key factor of an insulating fluid with ideal dielectric properties, which between voltage and current has a phase angle of 90°. Due to the existence of some contaminants, certain current flows through insulating system and in such case actual angle is less than 90°, an increase in tan delta is an indication of contamination of oil, and this may be due to oxidation of oil on aging. Polar components highly influence tan δ and is therefore it is very complex parameter. The dielectric dissipation factor of oil occurs when an insulating material is located between live parts and grounded part of an electrical circuit, leading to leakage in current flow. The insulation preferably leads to decrease voltage by 90° due to the dielectric nature of an insulating material, thereby making a rapid change in voltage between live and ground parts. Actually, no insulating materials are perfect dielectric in nature. Due to defect in insulating materials and their dielectric properties, tan δ of transformer oil is slightly less than 90° angle [8]. Cygan and Laghari [9] illustrated that with aging, increase in impurities (delta angle will be less than 90°) of transformer oil leads to increase in current flow through dielectric material thereby decreasing dielectric resistance as shown in Fig. 7.2.

Oxidation reaction in the presence of catalyst, temperature, and gasses leads to increase in polarity of oil. A variety of by-products are formed, some of which react further forming acids, sludge, and suspended particles. Short-chain acids are particularly detrimental because they speed up the aging of solid insulation [10]. The DDF is a measurement of polarity of molecules in oil, and vegetable oil cannot be compared directly with mineral oil, because polarity nature of triglyceride molecules in vegetable oil DDF is slightly higher than mineral oil. Rise in DDF in mineral oil indicates that formation of acids is due to intermediate species. However, higher

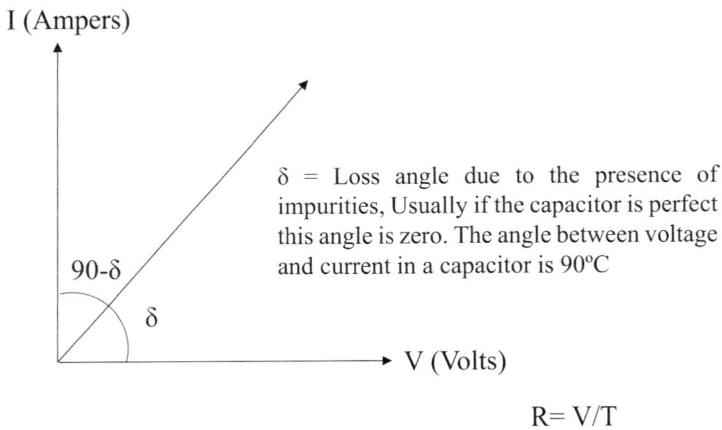

Fig. 7.2 Illustration of dielectric dissipation factor [10]

DDF of a vegetable oil varies by its internal ester linkages, not by intermediate species. Vegetable oil undergoes oxidation reaction and further increases its acidity, resulting to rise in DDF. Therefore, a condition monitoring management should focus on the rate of change in DDF instead of its absolute value.

7.2.1.3 Specific Resistance

Specific resistance (Ω) refers to obstruction to passage of an electric current through an electrical conductor [11]. Apart from superconductors that show zero resistance, a conductor with definite cross-sectional area shows a resistance directly proportional to its resistivity and length and inversely proportional to its cross-sectional area with other materials that show some resistance at any given point of time. A conducting material and temperature significantly influence its specific resistance.

Resistance (R) of a conductor is defined as ratio between voltage (V) and current (I) as shown in eq. (7.1):

$$R = V / I \tag{7.1}$$

where "V" is voltage and "I" is current that are proportional to a variety of materials.

At a definite temperature, resistance of a transformer oil between two sides of $1 cm^3$ block is defined as specific resistance, and measures in Ohm-cm; on increasing temperature, resistivity of an insulating oil significantly decrease.

7.2.1.4 Gassing Tendency

Gassing tendency of an insulating oil is defined as a chemical decomposition of certain hydrocarbon chains under the influence of thermal and electrical stresses. On decomposition of hydrocarbon chains, a small fraction of chemically reactive large free radical and gas evolution takes place. For example, molecular weight of this hydrocarbon chain before decomposition is in the range of 250–300, and losing a hydrogen atom or free radical (methyl) after decomposition leads to decrease in molecular weight by 10%. The crash between such large free radicals generates colloidal deterioration products with an average molecular weight of 450–550 that are non-soluble in oil. The gassing tendency of a liquid insulator is measured as an indication of a possible risk in operational safety of transformers.

The basic characteristic of gassing tendency of oil is a measure of change in volume under thermal and electrical stresses. There is unswerving relationship between certain molecules in transformer oil and gas absorption capacity. Chemical structure of vegetable esters is entirely different from mineral oil, and hence properties cannot be correlated with each other. Currently, no separate tests are available to evaluate the properties of natural ester oils; however, certain properties used to evaluate mineral are available in the literature [12]. The formation of sludge and degradation products resulting from degradation of natural esters, relationship between the gassing tendency, and its physicochemical properties have to be established.

Failure of liquid-filled transformer can be triggered by release bubble and was first experienced in oil-filled cables during the early 1930s [13]. The creation of a gas bubble in a liquid under electrical stress leads to formation of stressed liquid-gas boundary [14]. There is a direct relationship between certain molecules in the fluid and its capacity to absorb gas [15]. If the field stress is high enough, salient discharge, or non-troublemaking takes place with ionization and molecular activation of gas and liquid molecules. Such discharge does not go together at higher temperature; therefore, reactions are due to variation in frequency of applied voltage, electrical stress, pressure, nature of gas phase, electrons and chemical composition, and physicochemical reactions initiated by atomized gas phase.

Generally, under electrical and thermal stress, natural esters have basically lower gassing tendency (higher gas absorbing) than mineral oils [16]. Due to higher gas-absorbing capability of natural esters, many power industries are concentrating on using vegetable fluids in nonbreathing transformer application. Stray gassing is an unanticipated gas formed from certain fluids at a fairly low operative temperatures of a transformer ranging from 80 to 250 °C. These are not measured as an error or a problem with transformer. Dielectric fluids with low gas concentrations are considered to be less stray gassing.

Table 7.3 Chemical properties of natural esters vs mineral oil [4]

Test parameters	Specifications	Mineral oil	Natural ester fluid
Water content, mg/kg	< 200	15.0	20.7
Neutralization value, mg of KOH/g of oil	< 0.06	0.01	0.08
Corrosive sulfur	Non-corrosive	Non-corrosive	Non-corrosive

7.2.2 *Chemical Properties*

Chemical properties of an insulating liquid are one of the most important parameters that affect the performance of transformer in the long run. Recently, it is progressively more important that dielectric fluids give better balance between high functional recitals and less ecological impacts, and inside transformers it should be chemically and thermally stable and have superior insulating properties. Chemical properties of insulating oil such as water content, acidity, oxidation stability, aging characteristics, and corrosive sulfur are very important. Table 7.3 shows the important chemical properties of different liquid insulators and a detailed discussion is as follows.

7.2.2.1 Water Content

Water content of liquid insulator is an adverse pollutant that affects the performance of a transformer. Presence of higher content of water leads to reduction in the viscosity of oil further tending to reduce the boiling point than actual. Water content of vegetable oils is significantly different from that of mineral oil and is proportional to aging of the oil. A high water-in-oil solubility limit of vegetable oils is a great benefit of these fluids. Vegetable esters are capable of taking substantially more water than mineral oil and, when in contact with paper insulation, have an outstanding capability to dry paper insulator by absorbing moisture present in it. Ester fluids under elevated temperatures can experience hydrolysis, consuming available water from paper and improving paper aging characteristics. Water saturation of mineral oil is about 60 mg/kg at room temperature, [17] whereas vegetable esters have about 1000 mg/kg [18]. Dielectric strength of a liquid insulator starts to lower when virtual saturation enhances to about 40–50%.

The kraft paper winding in transformer is also an insulator in oil-filled transformers and it also badly affects with water content. On aging the transformer oil, water content gradually soaks solid insulation which decreases its insulating properties, thereby decreasing performance and efficiency of a transformer. In working condition, temperature of a transformer oil increases, and soaked paper again releases moisture to transformer oil undergoing oxidation reaction and further rises in neutralization value and moisture leading to reduction in quality of oil due to decomposition [11].

7.2.2.2 Neutralization Value

One of the foremost parameters for liquid insulator is neutralization value (acid number/total acidity). Neutralization value is a measure of acidity that refers to the number of milligrams of potassium hydroxide required to neutralize the H^+ ions present in 1 gram of oil. The purity of the processed oil can be evaluated by determining the total acid value.

Monitoring of acidity during the service is a key factor to ensure the safe operation and performance of the transformer. There are very few reports available on acidity studies of vegetable ester oils. Stability to oxidation of a liquid insulator can be analyzed by determining the neutralization value, dielectric dissipation factor, and specific resistance. Vegetable esters undergo oxidation and also decompose by hydrolysis leading to the formation of by-products (acids and alcohols). Hence, determination of acidity looks relevant, currently the standard test method is not available for natural esters, and presently available test method (IEC 62021–1) for mineral oils has been used for vegetable esters.

Ashraful et al. [19] reported that the neutralization value of natural esters increased proportionally with time. During storage at room temperature for 12 weeks, acidity for methyl esters (Jatropha, coconut oil, and pure palm oil) ranges from 1.0 to 3.0 mg of KOH/g and similar results were also obtained by Obadiah et al. [20]. According to ASTM D4625 (30 °C/50 weeks), they found the total acidity of Pongamia biodiesel rose up to 6 mg KOH/g.

7.2.2.3 Oxidation Stability

In both natural esters and mineral oils, oxidation process is a concern, but for several reasons, natural esters are not suitable for use in free-breathing transformers due to ultimate change in viscosity and should be separated from atmospheric contact and do not produce sludge in transformers. Mineral oils undergo oxidation process during service leading to the formation of sludge, but addition of antioxidant additives can delay oxidation reaction.

The degradation reactions that occur in the mineral oil are oxidation, dehydrogenation, and the cracking reaction (which leads to C-C bonding fracture, with the formation of alkenes). Oxidation and hydrolysis are the degradation reactions which occur in vegetable oil, leading to formation of CO, CO_2, alcohols, aldehydes, ketones, water, acids, etc. and also contain the degradation products resulted from degradation of the solid insulation (CO, CO_2, water, and furans).

Vegetable-based esters show low oxidation stability than their counterparts (mineral oil). However, method of oxidizing of natural esters differs from mineral oil. The oxidation by-products in vegetable fluids do not form any solid precipitates. Instead, oil starts to rise in the viscosity and eventually undergoes polymerization which is seen in high surface area to volume conditions.

The liquid insulators that are used in transformer must show higher stability towards oxidation. Oxygen present in oil and oxygen from the environment creates one or more influential factors (aging of oil by oxidation). The oxidation stability of oil is more significant if the oil is used in free-breathing transformer. Sulfur complexes present in mineral oils act as natural antioxidants, but synthetic antioxidant (2, 6-di tertiary butyl para cresol (DBPC)) can dope to the oil in order to slow down the degradation process. Though ester oil suppliers can dope with antioxidant additives, natural esters show lower oxidation stability due to more susceptible to biodegradation than mineral oil, leading the manufacturers to use a sealed design transformer. Hence, it is recommended not to use natural esters in free-breathing designed transformers because of ester oxidation [21, 22].

Properties of vegetable-based esters can be balanced by choosing proper composition of saturated and unsaturated triglycerides (TGs) during the formulation. However, double bonds being more delicate to oxidation, unsaturated TGs are less stable for oxidation than saturated ones, but on the other hand, saturated TGs show high pour point and viscosity. All these contemplations are valuable to select the best vegetable esters for electrical applications.

Early in the development of vegetable oils, various deliberations took place about the oxidation stability. The historically held opinion is that stability of vegetable oils for oxidation was inadequate and hence excluded them from considering as an insulating liquid. Some attempts were made to formulate vegetable ester liquids meeting the oxidation stability requirements of mineral oil by blending the vegetable esters with less eco-friendly synthetic esters. Attempts were also made to combine vegetable oils modified to have high olefin content with an increased amount of antioxidant additives.

Oxidation stability of vegetable transformer fluids must may be improved by adding antioxidant additives. The oxidation process occurs via free radical mechanism where antioxidants act as radical scavengers breaking propagation step of the process. Stability to oxidation of natural esters depends on distribution of fatty acids, refining process, and presence of natural antioxidants. It is also familiar that oxidation stability of vegetable oils decreases with increase in unsaturated fatty acid content. Oxidation stability of oleic acid (one double bond) is reported ten times greater than oxidation stability of linoleic acid (two double bonds) that is two times more stable than linolenic acid (three double bonds). Oxidation stability of natural ester oils also depends on refining process that consists of several steps including reduction in concentration of free fatty acids, waxes, metals, coloring pigments, and odors and also contribute to reduction in natural antioxidants concentration.

7.2.2.4 Aging Characteristics

Aging of oil that leads to deterioration of insulation is a function of time, temperature, moisture content, and oxygen content. Since modern preservation systems minimize moisture and oxygen contribution to insulation deterioration, temperature becomes most important controlling parameter of accelerated aging. Accelerated aging of insulating oils uses an oven to facilitate heat transfer to a container, real oil-filled transformer leading to overload need to alter internal characteristics of oil.

Oil-filled transformers contain liquid insulator used as an insulating and cooling media, the solid insulation, iron core, and copper. All these components affect the degradation of the transformer oil.

Accelerated aging studies that were performed by Mc Shane [23] using mass production transformers and was observed visually less fluid and other components decomposition in transformers using vegetable esters compared to transformers using mineral oil as an insulating fluid [23]. In a study by Choi et al. [24], it was observed that total acid number in the aged vegetable oil was higher than that of mineral oil. However, acids formed seem to be not harmful to solid or liquid insulation. The aged vegetable oil contains long-chain fatty acids, and presence of these acids in vegetable esters is non-corrosive compared to short-chain organic acids present in conventional mineral oil [24]. Therefore, neutralization value in vegetable esters does not cause any change in dielectric strength.

7.2.2.5 Corrosive Sulfur

It is generally accepted that the presence of corrosive sulfur compounds and the problems they cause in transformers are inherently an serious issue for mineral oil. Corrosive sulfur is not an issue for fluids other than mineral oil. Natural esters are produced from the seeds of crops and also do not contain sulfur that causes corrosivenes to metal components in the transformer. Hence, these fluids will not cause corrosive problems in transformers or other equipment [25]. Corrosive sulfur compounds are not naturally present in vegetable oils or other natural esters. Corrosive tests can verify that additives are non-corrosive and cross contamination with potentially corrosive oils has not occurred. Trace quantities of sulfur in transformer oil may be detected by sophisticated analytical techniques which will be chemically different from the types of sulfur compounds which react with metals. The manufacturers of vegetable oils should be able to provide evidence that their fluids are non-corrosive.

Existing methods anticipated that presence of corrosive sulfur components in transformer oils exists in as received state or extracted from gaskets, water-based glues, copper, and paper insulation (solid components) used during manufacturing of transformers. Sulfur can also be introduced into transformer through accidental means such as through usage of incompatible hoses [26]. In most cases, sulfur exists as part of an organic complex and the most reactive form of sulfur is its elemental form. Most of the sulfur compounds found to be stable and do not contribute to existing reactive sulfur compounds; however, by proper selection of temperature and ideal conditions, these stable compounds can be transformed into reactive compounds.

In a study, Rapp et al. [27] compared the properties of transformer filled with corrosive sulfur polluted oil and a sulfur-free vegetable fluids. Addition of passivators to retro filling of vegetable oil reduces the influence on corrosive sulfur. It was confirmed from a study that vegetable oil fluid doped with passivators retained the properties as per ASTM D6871 limits.

Table 7.4 Comparison of physical properties of natural ester with mineral oil [4]

Physical properties	Mineral oil	Natural ester fluid
Viscosity, cSt		
@ 40 °C	7.800	35.12
@ 100 °C	2.240	8.010
Density, @ 25 °Cg/cc	0.855	0.924
Pour point, °C	–40	–21
Flash point, °C	145	275
Interfacial tension, mN/M	40	30

7.2.3 *Physical Properties*

Vegetable and mineral oils considerably differ in their physical properties– especially, viscosity, density, pour point, interfacial tension and flash point as shown in Table 7.4. Unlike mineral oils, vegetable oils may not have much applicability in certain physical characteristics such as aniline point and interfacial tension. For example, aniline point is affected by the aromatics present in mineral oil of which natural esters have none in the refined form. Vegetable oils are hydrophilic in nature hence interfacial tension may not contribute to change quality of oil as highest interfacial values are in the range of 20 to 30 mN/m.

7.2.3.1 Density

Density is an essential physical property that can use in combination with other characteristics in several chemical industry applications viz., reactors for breaking of fatty acids or conversion of fatty acids into their derivatives, distillation for separation of fatty acids. Measurement of density or relative density of vegetable esters and its products is essential for conversion of measured volume to mass at a standard temperature of 20 °C. The density of natural esters shall be measured in accordance with ISO 3675 (reference -international method).

Density and viscosity play a significant role in the assessment of liquid insulators, there are very few reports published on optimization of the temperature at which the vegetable oils have improved performance in power equipment. Hence, it would be really helpful to know the importance of these characteristics within the wide temperature range to find out the most suitable temperature range for each vegetable oil. In an attempt, Forma et al. [28] studied the density behavior of various vegetable oils at different temperatures as shown in Table 7.5.

7.2.3.2 Kinematic Viscosity

Viscosity of an oil is defined as a measure of resistance to flow. Kinematic viscosity (KV) is the most prominent factor for dissipation or transfer of heat. Viscosity of an insulating fluid affects the capability to transfer heat by conduction, hence, cooling

Table 7.5 Density measurements of various vegetable oils [28, 29]

	Density, g/cc			
Temperature °C	Soybean oil	Coconut oil	Corn oil	Rapeseed oil
23.9	0.9193	–	0.9188	0.9078
37.8	0.9082	0.9107	0.9082	0.8977
48.9	0.9023	0.9033	0.9028	0.8898
60.0	0.8939	0.8949	0.8939	0.8829
82.2	0.8795	0.8795	0.8800	0.8681
100.0	0.8674	0.8669	0.8679	0.8564
110.0	0.8615	0.8695	0.8610	0.8501
25.0	0.915–0.918	0.916–0.918	0.915–0.917	–

the transformer by conduction is the main heat elimination mechanism. Higher viscosity of a fluid is likely to result in higher hot spot temperatures within the transformer. Rycroft [30] reported that the use of nature esters in power equipment leads to increase temperature by 1–3 °C.

Generally, at normal operating temperature of a transformer, vegetable oils show higher viscosity than mineral oils and less than silicone oils. The published reports reveal that transfer of heat in a transformer by convection is less efficient with vegetable esters. Viscosity is not a critical issue, but care has to be taken during the design of cooling system to power transformer. It should be noted that natural esters have a viscosity similar to mineral oil and, hence, minimize heat dissipation efficiency.

Higher kinematic viscosity may lead to a conclusion that cooling performance of esters is inferior to mineral oil, but this is not necessarily true. Because of high specific heat capacity and higher thermal conductivity, natural esters is able to transfer higher amount of energy. Therefore, cooling performance over the lifetime of a transformer can be an advantage if the cooling design is adjusted to the characteristics of natural ester.

Kinematic viscosity strongly influences impregnation of cellulose-based solid insulation. In normal processing, temperature range (about 70 °C) viscosity of natural ester fluids is four times more than that of mineral oil. Therefore, mechanical design of large insulation structures as well transformer impregnation process has to be adjusted as per requirement. For example, design of large wooden and pressboard elements needs to be reconsidered while using esters as an insulating medium. Installation of drying and impregnation holes is suggested to speed up impregnation process. Impregnation tests of pressboard imply that it is necessary to extend impregnation time if the same boundary conditions (such as liquid temperature) are maintained. Due to oligomerization of fatty esters and unusual exposure to air and heat, viscosity of ester fluids may increase on oil aging. Literature does not demonstrate such evolution, and viscosity of ester composition remains unaffected even after two or more years of operation [31].

It was observed that with natural esters it takes at least two times higher to complete impregnation process of laminated pressboard. The liquid temperature for both specimens is same; natural and synthetic esters behave similarly because of similar viscosity behavior. In order to speed up the process of impregnation, ester liquid should be processed at slightly higher temperatures. However, liquid temperature has to be limited depending on the composition of natural ester. Especially in case of natural esters, additives will be consumed at a higher temperature in vacuum condition. To balance the process for successful transformer application, it is advisable to extend the time of impregnation and slightly increase temperature of ester liquid. This combination will achieve a sufficient impregnation of solid insulation material.

7.2.3.3 Pour Point

The temperature at which fluid just flows under specific conditions is defined as pour point. Pour point is an important parameter that decides the performance of transformer oils at lower temperatures, especially when it is required to cold start a transformer under very low-temperature conditions. Vegetable esters have pour point in the range of −15 to −25 °C [32], which is higher than that of mineral oil, but studies have shown that significant cooling starts down to −30 °C. Addition of pour point depressants to natural ester fluids can be atleast similar to typical mineral oil.

Insulating fluids used in power transformer act as both coolant and insulate internal constituents and are expected to sustain acceptable flow during extreme temperature conditions (hot or cold). Additives (cold flow modifiers or pour point depressants (PPD's)) are doped to natural ester-based fluids and high molecular weight hydrocarbon (HMWH) fluids to enhance the performance at very low temperatures. Lubricants (see Chaps. 3 and 4) and co-solvents have higher levels of PPDs and can doped in vegetable ester fluids [33–35]. Analysis of natural ester blends shows some pour point improvement. However, under continuous cold conditions, they lose this property.

In some global regions, there are special environmental conditions, especially with respect to a temperature profile. In case of natural esters, low-temperature application can be critical because esters have higher pour point than mineral oils (about −21 °C.). However, solidification of esters begins at approximately −10 °C, so it is important to consider this to design transformer. Many transformer specifications require lowest ambient temperatures of less than −25 °C [36].

Transformers used in outdoor applications may expose prolonged time at low temperatures (sub-zero) (Himachal Pradesh, Jammu, and Kashmir). The role of energized transformers is to generate sufficient heat to maintain free flowing dielectric liquid. However, idle equipment or outdoor spares may go weeks or months stored at extremely low temperatures. During these conditions, a vegetable ester fluid may vary between solid and liquid states forming a typical type of sediment.

Acid chain length, unsaturation, and branching are the structural properties through which pour point of a vegetable ester can control. Both unsaturation and branching in fatty acid have positive influence in lowering pour point. Unsaturated fatty acids are more effective in reducing pour point than branched fatty acids of similar carbon chain length. The presence of aromatic groups in vegetable esters also has a positive influence in keeping low pour point.

7.2.3.4 Flash and Fire Point

The foremost advantage of vegetable oils is their comparatively greater fire and flash points than currently available mineral oils. A discrete margin of safety is qualified to insulating fluids with higher flash and fire points. Fire point plays a key role during transportation of oils and installation of transformers in indoor or in commercial buildings with or without supplementary fire safeguards. Vegetable esters have fire points of about 360 °C (see Tables 7.1 and 1.8) and qualify as "K" class as per IEC 61100. They have been used in practical installation and have regulatory advantages in many sites [37–41].

7.2.4 Miscellaneous Properties

7.2.4.1 Heat Transfer

Liquid insulators have to ensure cooling mechanism of transformer in order to maintain acceptable temperature in addition to dielectric properties, thus reducing aging of oil. Kinematic viscosity, specific heat capacity, expansion coefficient, and thermal conductivity of a transformer fluid decide the removal of heat capacity from coils toward outside the transformer.

Liquid insulators in a transformer must confirm the dissipation of heat and is realized by both thermal conductivity and convection [42]. The convection represents all properties, viz., viscosity, specific heat, and thermal expansion that leads to heat transfer by fluid displacement (coefficient), whereas conduction is understood within the fluid. In a recent study by Perrier et al. [43], it was observed that kinematic viscosity is the most effective parameter for transferring heat.

7.2.4.2 Thermal Properties

Vegetable oils show higher thermal stability than petroleum-based mineral oils. In natural ester fluids, thermal degradation of solid insulation is reduced by a factor of 5–8 times compared to mineral oil [44–46]. Increase in specific heat capacity of natural esters with increasing temperature is stronger than those of other conventional liquids. Thermal conductivity of vegetable oils is higher compared to mineral oil over a wide temperature range.

7.2.4.3 Environmental Safety and Biodegradability

Safety in terms of environment is ensured by two rudimentary criteria, viz., no or low toxicity and biodegradability. Chapter 2 illustrates complete details on biodegradation and toxic nature of vegetable oils. In general, fluids that possess a rapid biodecomposition rate and low toxicity are characterized as eco-friendly and are important at the time of considering to use in environmentally sensitive areas like water decontamination. The term "biodegradability" replicates the degree to which liquids break down by the action of microorganisms in soil or water passages during spillage or leakage. Clearly, it is an advantage if spilled fluids undergo degradation by natural mechanism avoiding the need to set up expensive cleanup procedures. The schematic representation of biodegradation process of vegetable esters is shown in Fig. 7.3.

In both aerobic and anaerobic atmospheres, natural esters have better tendency for biodegradation than mineral oil, reducing cleanup costs on spillage. As per available literature, natural ester oils undergo about 70–100% biodegradation in a period of 28 days (see Tables 2.1, 2.2, 2.3, and 2.4) [47, 48]. Degradation of vegetable esters has both an advantage and disadvantage, while oil spills to environment degradability process help in decomposing oil in a safe way, whereas oils undergo degradation by the action of oxygen so that oil must be prevented from degrading inside the transformer.

The ecological characteristics of natural esters are excellent; they undergo biodegradation quite rapidly and completely and produce no toxic products. Vegetable-based insulating fluids do not contain semi-volatile organics, dangerous aromatics,

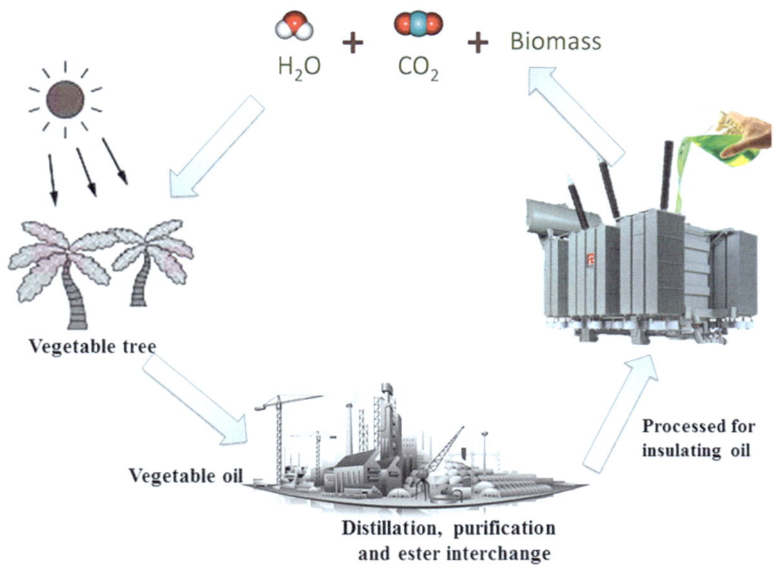

Fig. 7.3 Schematic representation of green cycle of vegetable oils

volatiles, halogens, etc. The high degree of biodegradability of esters enables a financial benefit for facility investment because no separate containment facility needs to be installed to prevent pollution of the atmosphere in the event of leakage.

Biodegradation is a mechanism by which organic substances degrade to low molecular weight compounds by enzymes produced by action of microorganisms. Organic material undergoes degradation both in the presence of oxygen (aerobic) and the absence of oxygen (anaerobic). The process of conversion of organic matter into minerals is known as biomineralization [49]. There are two types of biodegradation. First is primary biodegradation, known as transformation of a substance by the action of microorganisms such that a change is initiated in some specific measurable properties of the substance that alters physical characteristics of a compound while leaving molecule largely intact [50]. Intermediary metabolites produced may, however, be more toxic than original substrate [3]. Second is ultimate or complete biodegradation achieved when a substance is totally utilized by microorganisms resulting in the formation of low molecular weight by-products (carbon dioxide, methane, water, minerals, and new microbial cellular components which is referred as mineralization [51].

There are different test methods for biodegradation studies of matter which measure loss of dissolved organic carbon substances (water soluble) and uptake of oxygen due to the action of microorganisms (BOD) and amount of carbon dioxide (or methane in anaerobic process) produced during a specified period; and there are yet other measures, the uptake of oxygen by means of microorganism activity (BOD) [52].

7.3 Additives for Vegetable Fluids

Additives are chemical compounds that are doped in liquid insulators to improve the properties without affecting the actual properties. Natural esters comprise additive packages to reduce pour point and help to increase stability to oxidation in the presence of catalyst and temperature, and in other instances, they have copper deactivators and antimicrobial agent that inhibit oxidation reaction, leading to increase in oxidation stability. In contrast, mineral oil has either no additives or merely oxidation inhibitors. It is not known if any opposing characteristics exist in transformers during the long-term usage of vegetable oils.

7.3.1 Antioxidant Additives

Molecules capable of reducing or inhibiting oxidation reaction of other molecules are known as antioxidants (oxidation inhibitors) [53]. Oxidation reaction in organic systems produces free radicals, which commence chain reactions. Doping of oxidation inhibitors terminates chain reaction by eliminating free radical intermediates;

hence, most oxidation inhibitors often act as reducing agents. Generally, free radical leads to an oxidation reaction which extremely enhance the corrosion. Since these compounds remove free radicals, oxidation is reduced to maximum extent, thereby reducing corrosion. Oxidation inhibitors play a vital role as a reducing agent, free radical scavenger, singlet oxygen scavenger, and chelator [53].

The real fact is that in their natural form, vegetable esters are lacking in oxidation stability, which is the main drawback in using these fluids as an insulating fluid in transformer application. Oxidation stability of oil is defined as a measure of time taken to initiate oxidation process. In general, the structure of hydrocarbon, heteroatom concentration, oxygen concentration, and temperature affect the rate of reaction in oxidation process [54]. The more unsaturated fatty acid is involved, the greater the oxidation susceptibility. For instance, linolenic acid esters existing in soybean oil are particularly sensitive to even the least kind of oxidation [55]. Regarding storage ability, a high saturated fatty acid level confirms to be more beneficial compared to low unsaturated vegetable oils [54].

The propensity of an oil to associate with oxygen and become gummy is known as drying and is measured by iodine value, in other words, a measure of unsaturation level of oil (a higher iodine number indicates higher unsaturation) [56]. Toshiyuki et al. [57] studied oxidation stability of refined vegetable oils and found to be determinable by fatty acid composition. In general, it is seen that non-refined oils have better stability than refined ones [58] and is attributed to the fact that refining step removes certain amount of natural antioxidants in vegetable oils [59].

Corn oil shows better oxidation stability than soybean oil, while rapeseed oil gives better performance than olive oil [60]. Ester oils show the trend of superior stability in unrefined form than that of refined state. In a study, Isbell et al. [60] reported that meadowfoam oil is the most stable vegetable oil [60]. Sunflower oil and crude jojoba oil with high oleic content have good oxidation stability [60]. The existence of free fatty acids has a pro-oxidant effect on vegetable oils; hence, refining practices are important in the formulation of vegetable oils as a dielectric fluid [61]. Indecorous handling and manipulation of raw materials lead to enzymatic activity that produces free fatty acids [61]. Further investigations on manufacturing practices of natural esters also reveal the importance of solvent used for extraction. Solvents such as hexane or petroleum ether have a property of extracting only nonpolar species; isopropanol can extract some polar and high molecular weight compounds [62]. Natural antioxidants and pigments present in vegetable seeds lead to extend shelf life and superior oxidation stability [62].

Numerous experimental studies were established on the effect of antioxidants on stability of natural esters toward oxidation in electrical industry applications. An important class of oxidation inhibitors contains phenolic compounds, viz., butylated hydroxyanisole (BHA), butylated hydroxytoluene (BHT), tert-butylhydroquinone (TBHQ), and propyl gallate. The usage of ester fluids intended for domestic and industrial applications is common.

Various research investigations were held on determining the effect of natural antioxidants contained in red pepper oil added to sunflower and soybean oil. Results reveal that they provide protection against light-induced auto-oxidation. Inhibition effect on oxidation can be estimated by determining fatty acid formation and peroxide values [63, 64]. In a study by Emanuel and Lyaskovskaya inhibiting action of tocopherols on rapeseed and palm kernel oils was estimated by computing the presence of oxidation products, monoaldehyde indicating some measure of protection provided by these natural antioxidants [64]. The above study on inhibitive effect of natural antioxidants contained in red pepper oil revealed that phenolic antioxidant (butylated hydroxytoluene) shows more effective than natural antioxidant additives [63].

The consequence of phenolic oxidation inhibitors on enhancing oxidation stability of soybean oil is available in literature. Dunn [65] observed the oxidation stability by means of pressurized-differential scanning calorimetry (P-DSC). Improvement in oxidation stability for both static and dynamic conditions was observed, in addition to antioxidants, which include propyl gallate (PrG), BHA, BHT, and TBHQ. Dunn [65] further studied on the relative efficacy of various oxidation inhibitors on the properties of static and dynamic conditions. In addition, relationship between oxidation inhibitor content and oxidation stability was established. Interestingly, in a study Jung et al. [66] studied effect of various concentrations of tocopherol on oxidation stability of soybean oil is higher resulting in pro-oxidant effect [67]. General conclusion is that increased loading of antioxidants up to certain extent adds advantage in improving oxidation stability.

Ruger et al. [67] studied the capability of various antioxidants and chelators to interrupt increase in viscosity of soybean oil by auto-oxidation. Senthil Kumar [68] performed a study on effect of oxidation inhibitors on various characteristics of natural esters (house, neem, mustard, punna, and castor oils). It was reported that vegetable oils have shown superior properties than mineral oil viz., electric strength, flash point, fire point and viscosity. It was also observed that doping of oxidation inhibitors in natural ester fluids showed improved properties compare to natural esters in absence of oxidation inhibitors.

Current area of research interest, that are seen in owning good antioxidant properties is finding an effective replacement for conventional synthetic oxidation inhibitors from various extracts of plant species. Synthetic antioxidants having serious health effects on explosion to humans are known to contribute to liver enlargement and increase in microsomal activity [69, 70]. Inhibition of lipid peroxidation was found to be comparable to inhibitions obtained when treated with vitamin C and E [71, 72]. Using peroxide values as a measure, antioxidant effectiveness of methanol extracts was tested in sunflower oil stored at 70 °C. Efficacy in stabilizing sunflower was obtained from methanol extracts of *P. buggier,* and *S. Alexa.* These tests and their conclusions suggests strongly the probability of having a feasible source of natural oxidation inhibitors of superior performance [69].

7.3.1.1 Mechanism of Antioxidant Additives

The ability to oxidation of natural esters depends on extent of unsaturation in their Olefin components. As a whole, oxidation process in vegetable oil take place in presence of heat, metal catalyst etc., resulting in conversion of unsaturated particles to free radicals. These free radicals are readily oxidized to produce aldehydes, ketones, or acids (hydro peroxides and organic components) leading to adverse odor and flavor of rancid fats [73]. Radical induced chain reaction via initiation, propagation, and termination at an enhanced rate after initial induction period is known as auto-oxidation. Comparatively stable radicals formed by abstracting hydrogen atoms from allylic methylene groups in olefin complexes. Proposed mechanisms for free radical chain reactions are existing in the literature [74].

Antioxidant is used in a universal sense to denote any kind of chemical that constrains oxygen attack [75]. On addition of these materials to vegetable oils, oxidation reaction intrudes by specially reacting with free radicals to form a steady compound that does not rapidly react with oxygen [76]. Function of oxidation inhibitors is either by restricting the formation of free alkyl radicals in initiation step or by interposing propagation of free radical chain [77]. In shortening propagation step, inhibitors act as a hydrogen donor which leads towards elimination of diffusion of free radical chain.

The most widely used oxidation inhibitors are hydroxyl phenol components with different ring replacements and categorized by owning low activation energies for donation of hydrogen atom. They relatively undergoes reaction with lipid free radicals to form stable and complex components [77]. During investigation of phenolic inhibitors, it was found that anti-inhibition ability depends on number of phenolic clusters occupying 1, 4 or 1, 2 positions in an aromatic ring, volume and electronic characteristics of ring substituents [74]. In revealing the mechanism of oxidative inhibition, it is recognized that in antioxidants act as oxygen interceptors thereby terminating chain reaction to propagates process [78]. Conventional antioxidant activity involves hydrogen donor to free radicals leading to creation of complex between lipid radical and inhibitor radical formed due to loss of hydrogen. Schematic representation of antioxidant activity is shown in Scheme 7.1.

Scheme 7.1 Mechanism of chemical activity of antioxidant additives

$$R\bullet + AH \longrightarrow RH + A\bullet$$

$$RO\bullet + AH \longrightarrow ROH + A\bullet$$

$$ROO\bullet + AH \longrightarrow ROOH + A\bullet$$

$$R\bullet + A\bullet \longrightarrow RA$$

$$RO\bullet + A\bullet \longrightarrow ROA$$

$$ROO\bullet + A\bullet \longrightarrow ROOA$$

$$\text{Anti oxidant} + O2 \longrightarrow \text{Oxidized Antioxidant}$$

7.3.2 Pour Point Depressants

The ability of a fluid to flow under low-temperature, low-shear conditions is crucial to operate transformer expected to run in cold climates. Without proper selection and treat of a pour point depressant, insulating oil formulation exhibits poor low-temperature properties. General mechanism of pour point depressants in oil is shown in Fig. 4.6. Poly alkyl methacrylates are the first polymeric pour point depressants, continue to be viewed as a best chemistry available today. Primary reason for this widespread preference is molecular structure of polymers and tremendous flexibility in chemical structure.

Natural triglyceride ester fluids have tendency to delay transition from liquid to solid state that occurs at temperatures higher than measured pour point value and show significantly different low-temperature properties. These compounds have noteworthy low-temperature properties than mineral oils and synthetic esters. Natural ester fluids show comparatively high solidification temperatures even on doping with pour point depressing agents.

Triglycerides have poor low-temperature viscometric and tend to thicken at temperatures below −10 °C. Many industrial fluids have a pour point less than −25 °C. In order for triglycerides to be used successfully as industrial base fluids, their low-temperature viscosities must be improved. A number of complexes are known to improve low-temperature viscosities of vegetable oils. These compounds are known as "pour point depressants" (PPDs). Known PPDs for vegetable esters include, but are not limited to modified carboxyl based polymers; acrylate polymers; nitrogen based acrylate polymers; and methylene linked aromatic compounds. Unfortunately, known PPDs are not biodegradable. Therefore, advantage in low-temperature viscometer that is gained by using these PPD's is largely offset by decrease in biodegradability of a resulting product. Also, manufacturing and environmental specifications limits the total amount of non-biodegradable material that can be used as a particular industrial fluid. Biodegradable PPDs would meet applicable specifications and not compromise overall biodegradability of industrial fluids are greatly needed (see Sect. 3.4).

7.4 Performance and Evaluation of Vegetable oil as Insulating Fluids

Natural esters are triacylglycerides and vary chemically from mineral insulating oils (see Table 7.1) and hence viscosity of vegetable oils is relatively higher than mineral oil (see Table 7.5) and this is partly compensated by other thermal properties such that thermal design of a transformer is comparable for both mineral oil and natural esters. Difference in chemical composition reveals change in fluid properties, viz., viscosity, thermal expansion, thermal conductivity and heat capacity of a fluid can optimize in designing cooling system of a transformer.

Hermetical sealing of a power transformer is an appropriate measure to confirm smooth operation of transformer using eco-friendly vegetable ester fluids this is due to low oxidation stability of vegetable esters. Working life time of a transformer and insulation (solid and liquid) can be extended by combination of two measures viz., avoidance of contact between humidity and oil and choosing natural esters having higher ability of moisture saturation. Additional advantage with vegetable esters is option to overload a transformer for an extended shelf life without adverse effect on rate of aging.

Natural ester fluid dissolves water approximately 20 times more than a mineral oil at a given temperature [79]. Limits for water in mineral oil should therefore not be directly used for vegetable oil insulation. For instance, the effect of water on lowering the breakdown voltage of oil is more related to percent of saturation, rather than water concentration [80]. The water content of oil is usually expressed as either mg/kg (mass water/mass oil) or as water activity (vapor pressure of soluble moisture/vapor pressure of pure water at a particular temperature).

Water content of vegetable fluids significantly varied from mineral oil. At room temperature, water saturation of mineral oil is about 60 mg/kg whereas vegetable esters is approximately 1000 mg/kg [17, 18]. Breakdown voltage of a liquid insulator starts to reduce when relative saturation enhances to about 40–50%. Instead of using a percentage of saturation, absolute moisture content allows direct comparison between vegetable esters and mineral oil.

7.5 Challenges and Technical Difficulties

7.5.1 Challenges

The challenge of vegetable oils is enhancement in their performance (specifically oxidation stability and pour points) as an insulating fluid in power transformers. Advances are unavoidable and are previously noted with growing research emphases in these areas. Market and governing pressures to minimize accountability risk on exposure of transformers filled with mineral oils are growing. In addition, there are demands to improve equipment efficacies and adopt more eco-friendly options in power systems; hence, transformer industry has been emerging with new conceptions.

7.5.1.1 Fault Detection Using DGA Results for Alternative Oils

Dissolved gas analysis (DGA) is an effective technique to notice initial faults in mineral oil filled transformers and was used by various researchers. Data provided by DGA analysis is exceptionally important to asset managers with electrical industries. Fault gasses are formed when sufficient energy breaks chemical bonds to form reactive fragments. These fragments combine to form the range of fault gasses. Although chemical structure of mineral oil, ester fluids, and silicone fluids are basically different, they all contain a similar type of bonds, for example, C–H bonds, such type of fault gasses are similar for each of the fluids. For a given fault, amount and ratio of gasses may differ from fluid to fluid.

Liquid insulators undergo heating at fairly low temperatures (90–200 °C) and generate gasses is known as strays gassing [81]. On energizing transformer filled with vegetable esters at comparatively low temperatures, generation of significant amount of stray gasses like ethane and hydrogen are observed over a period (weeks to months) [82]. In case of FR3's (commercial grade natural esters) it was detected that vulnerability towards formation of stray gas differs from batch to batch [83]. In case of partial discharge, compare to mineral oil, vegetable esters generates higher amount of hydrogen and small amount of acetylene gases at the same applied voltage level [84].

Faults and stresses in transformers filled with vegetable fluids prone to the formation of noticeable amounts of fault gasses [85]. Compare to mineral oil, formation of gasses in natural esters is higher during partial discharge and lower in arcing. Gases generated due to thermal faults of a transformer filled with vegetable esters is sufficient to use for diagnostic methods. Due to the above reasons, it becomes obvious that for prevailing fault diagnostic interpretation systems limits for natural esters have to be framed accordingly.

Gassing properties is also one of the important features to be considered for condition monitoring of a transformer. Most commonly used technique to detect any emerging faults in a transformer is dissolved gas analysis (DGA). DGA assessment gives early warning to take preventive action and reduce economic losses. Method of DGA assessment for vegetable esters varies from that of mineral oil due to variation in key fault gases of vegetable esters [86, 87]. During operation of transformer, key fault gases are important for interpretation of faults in the transformer. Therefore, during service.it is crucial to identify key gases of natural esters under different types of faults so that correct elucidation and proper preventive action can take.

Using DGA, two broad categories of faults (thermal and electrical) can be detected. Electrical faults can further categorize into low energy partial discharges and high energy arcing, faults may occur in oil or in cellulose-oil interface. In order to apply DGA diagnostic method on transformers filled with vegetable oil, it is essential to determine if same type of fault gasses are generated, rate of generation and concentration of fault gasses in conventional mineral oils. Due to oxidation reaction of unsaturated fatty acids, vegetable esters develop higher concentrations of ethane gas than mineral oil, thus ethane can be observed as a key gas for interpretation of dissolved gas analysis of vegetable oils [88].

Recently, investigations on comparison of DGA between natural ester fluids and mineral oils have engrossed lots of attention. Martin et al. [83] investigated dissolved gases of normal service power transformers filled with commercial vegetable esters and mineral oil indicating hydrogen and ethane are evolved considerably in case of transformers filled with vegetable esters. Nevertheless, there is no much data on key fault gasses of natural esters, various dissolved gasses results obtained from the literature concludes that natural esters and mineral oil varies in dissolved gas analysis under thermal and electrical faults [87, 89–92].

However, previous studies reveal that more attention on comparing dissolved gasses of mineral and vegetable oils. Due to the fact that different esters were produced from various types and proportions of triglycerides, dissolved gas of different vegetable oils extracted from different source varies. Thus, there is a prerequisite to validate and develop diagnostic methods for DGA of vegetable oils. Moreover, it is essential to study gas generation mechanism of various natural esters that helps to deliver a hypothetical basis for conniving investigative techniques of natural ester insulators of varied chemical compositions.

7.5.1.2 Interaction Between Natural Esters and Cellulose Paper

The solid insulation (cellulose paper) in transformers unavoidably has some porous structure, to be infused with oil. Transformation of fluid in porous structure primarily depends on size of the capillaries, inside and outside pressure of pressboard block, and viscosity of fluid. Impregnation of thick composite pressboard blocks offers itself as one of the technical challenges in vegetable oil filled transformers, and impregnation procedure needs to be developed for supervision of manufacturer. Vegetable ester fluids have higher viscosity than mineral oil, hence, it is presumed that longer time is essential for esters to infuse paper insulation. Infusing time of oil is a function of viscosity, surface tension and capillary effect of oil inside paper insulation. Dai and Wang [93] proved that effective impregnation of solid insulation with natural esters can be achieved within same period by increasing temperature by about 60 °C.

Paper insulation in a transformer normally transfer electrical stress between turn to turn under AC operating voltage and impulse surges. The designed electric stresses onto paper insulation need to be lower than dielectric strength and a proper safety margin should be maintained. Breakdown voltage of oil impregnated paper decreases as the thickness of paper increases and within a controlled moisture level ester impregnated paper shows analogous BDV to that of mineral oil impregnated paper. Similar to pressboard, paper with less density would have lower AC breakdown voltages [94]. Laboratory test results showed that layer insulation paper with a density of 0.75 g/cm^3 has lower dielectric strength than paper with a density of 0.93 g/cm^3. Nevertheless, test results reveals that ester impregnated layer insulation paper has comparable dielectric strength to that of mineral oil impregnated paper.

The high water uptake capacity of vegetable fluids absorbs moisture from paper, thereby minimizing hydrolytic deterioration. Thus, improves thermal insulation life of a cellulose paper, life of solid insulation in natural ester fluid is five to eight times greater than that in mineral oil at 50% retained tensile strength [95]. At voltage level of 132 kV and above, insulation system of transformer *contains* oil and oil-impregnated cellulose, and shelf life of a transformer is mainly dominated by cellulose insulation. To apply esters in large transformers, ester impregnated solid insulation should be proven to have *relatively higher* dielectric strength of mineral oil impregnated solid insulation.

7.5.1.3 Impregnation of Solid Insulation with Ester Fluids

Solid insulation (paper, pressboard, and blocks) needs to be impregnated by insulating oil to have better dielectric properties. There are three variables that would affect impregnation result: pressure, impregnation time under vacuum and viscosity of a fluid. Ester based fluids have higher viscosities than conventional mineral oil, which raises an issue whether the impregnation by ester would take much longer time than mineral oil.

Impregnation of both solid insulation and pressboard brings no technical problem due to their high thickness. Experimental tests in laboratory indicates that paper with a thickness less than 0.5 mm can be fully impregnated by an ester fluid under vacuum of 15 mbar at 60^0 C within 12 h, whereas 3 mm thick pressboard would need 48 h. Impregnation of laminated blocks, either pressboard blocks or wood blocks brings more engineering challenges because of their higher thickness, the more viscous fluid and consequently a longer impregnation time. Some comparative laboratory experiments were carried out in the laboratory to study impregnation process of blocks by mineral oil and ester fluids. Increasing the temperature of oil can reduce its viscosity and thereby shorten impregnation time, still the exact temperature and time need to be determined for transformer manufacturer.

7.5.2 Technical Issues

It is indispensable to use design review method to recognize formerly established design prerequisite justification and/or new progress. It is true specifically for reactors and transformers that are to be filled with vegetable fluids. While, basic design and construction is comparable to previous designs, a key component was substituted, i.e. liquid insulator. "Oil-immersed" transformers and reactors rely on the "oil" to provide part of insulation structure as well as allow thermal losses. It is essential to confirm its role with regard to both manufacturing and design, if this liquid insulator is to be altered. In the above cases, design and evaluation procedure, it is emphasized on the difference between two liquids that is desirable to understand in case their impact was noteworthy during manufacturing, operation and design.

7.5.3 Dielectric Issues

The vegetable esters used in oil-immersed transformers and reactors have a similar relative permittivity of kraft paper and pressboard. Different voltage distributions under transient conditions such as impulse application (lightning strike) produced due to capacitance of change in insulation structure. The capacitance of insulation structure varies and leads to dissimilar voltage distributions under transient conditions such as impulse application (lightning strike). Voltage stress distribution with insulation structure also changes, such that for a given voltage distribution, voltage stress in liquid was less for vegetable esters than mineral oil. This is advantageous and permitted to upper levels of voltage to endure for definite electrode and insulation structures. During manufacturing and service itself, information about altered acceptance norms for condition of oil should be available for use. Transformer and shunt reactor, both filled with vegetable fluids are effectively conceded all their dielectric tests.

7.5.4 *Thermal Issues*

At a particular temperature, vegetables fluids have higher viscosities than mineral oil. This decreases flow rate of a liquid for a given dynamic head that leads to higher temperature variance between top and bottom of the cooler. This is noteworthy for naturally cooled transformers, although mean oil temperature rise is controlled by cooler, top oil is controlled by natural thermosyphon fluid flow and is effectually summation of mean of oil rise and half the top to bottom fluid rise. Where the cooling system uses forced directed fluid flow, effect is minimal, providing that precise rating of pump is used to consider more impedance to oil flow.

Literature indicates higher temperature rise takes place in case of transformers filled with vegetable esters compare to mineral oil [96] suggesting that lifespan of natural ester filled transformer is lower than equipment filled with mineral oil. However, in presence of natural oil impregnated paper decomposes much slower [97], leading to permit over loading transformer at a given level and maintaining same lifespan as that of mineral oil filled transformers [97–100].

7.6 Conclusion

Todays researchers and industrialists around the world are willing to shift from petroleum based mineral oils to natural esters (renewable and eco-friendly). These renewable resources have superior insulating properties and are well-suited to use without any environmental hazard (animals and plants). Transformers filled with mineral insulating oils facing problem in many aspects, hence, both in market and regulatory pressures increasing to reduce liability risk on exposure of mineral oil filled transformers (Distribution and power). In addition, there is a scope to increase efficacy of equipment and implement more ecological possibilities in power equipment. In view of this paradigm, transformer manufacturing companies have been emerging to new ideas. Additional investigations desirable related to long-term features and ability to pleasant the environment.

Development of ester insulating fluids allows alternative to mineral oils. In the domain of distribution transformers, practices shows that performance of natural esters is adequate. The familiarity with natural oil fluids for power transformers is still limited, but however positive superior fire properties and optimistic conservational performances have been well documented. Tuning of vegetable fluids perhaps speed up the execution of such liquids in power transformers. Mineral oils is cheaper than vegetable oils, although cost is more, improved fire safety and biodegradability of vegetable fluids may lead to positive decision to use synthetic or natural esters in the future.

References

1. Darwin A, Perrier C, Foliot P (2007) The use of natural ester fluids in transformers. In Proceedings of MATPOST conference, Lyon (France), Paper 0036 (November 15–16, 2007)
2. Experiences in service with new insulating liquids, working group Cigre A2–35
3. Thomas A. Prevost Dielectric Properties of Natural Esters and their Influence on Transformer Insulation System Design and Performance.
4. 2nd International Conference on Innovations in Electrical and Civil Engineering ICIECE'2013) Dec. 17–18, 2013 Pattaya (Thailand)
5. Darwin A, Perrier C, Folliot P, The Use of natural ester fluids in transformers, MAT Post'07
6. Vukovic D, Tenbohlen S, Report on Comparative evaluation of breakdown strength of natural esters and mineral oil
7. Martin D, Wang ZD, Dyer P, Darwin AW, James IRA (2007) Comparative Study of the Dielectric Strength of Ester Impregnated Cellulose for Use in Large Power Transformers, International Conference on Solid Dielectrics, Winchester, July 8–13
8. Accessed in November 2013.http://electrical-engineering-portal.com/resources/knowledge/electrical -machines/dielectric-dissipation-factor
9. Cygan S, Laghari JR (1987) Dependence of the electric strength on thickness, area and the volume of polypropylene. IEEE Trans on Elec Insul EI-22(6):835–837
10. Lelekakis N, Martin D, Wijaya J (2012) Ageing rate of paper insulation used in power transformers part 2: oil/paper system with medium and high oxygen concentration. IEEE Trans on Dielectrics and Electrical Insulation 19:2009–2018
11. Accessed in October 2013. http://www.electrical4u.com/transformer-insulating-oil-and-Types-of-transformer-oil/
12. Lewand L, (2004) Laboratory Testing of Natural Ester Dielectric Liquids. Neta World, pp 1–4
13. Pahlavanpour B, Elkund M, (2004) Gassing properties of insulating liquids. Electra, No. 214
14. Arakelian VG, Fofana I (2009) Physicochemical aspects of gassing of insulating liquids under electrical stress. IEEE Electr Insul Mag 25(3):43–51
15. Lewand L and Griffin P (2005) Gassing Characteristics of Transformer Oil under Thermal Stress. Neta World, pp 1–4
16. Davydov, VG, (June/July'2014) Consultant, Australia, Natural Ester Dielectric Liquids for Transformers and Other HV Apparatus
17. IEEE guide for acceptance and maintenance of insulating oil in equipment, C57.106-(2002) Institute of Electrical and Electronics Engineers, Piscataway
18. Lewand L. (2001) Laboratory evaluation of several synthetic and agricultural-based dielectric liquids. Doble International Client Conference, 5E, Boston
19. Ashraful AM, Masjuki HH, Kalam MA, Rahman SMA, Habibullah M, Syazwan M (2014) Study of the effect of storage time on the oxidation and thermal stability of various biodiesels and their blends. Energy Fuel 28(2):1081–1089
20. Obadiah A, Kannan R, Ramasubbu A, Kumar SV (2012) Studies on the effect of antioxidants on the long-term storage and oxidation stability of Pongamia pinnata (L.) Pierre biodiesel. Fuel Process Technol 99:56–63

21. Base oil handbook. Nynas
22. Martin D, Wang ZD, Darwin AW, James I (2006) A comparative study of the chemical stability of esters for use in large power transformers. Electrical Insulation, and Dielectric Phenomena, 2006 IEEE Conference on, 15–18 Oct. 2006
23. Mc Shane CP, Gauger GA, Luksich J (1999) Fire resistant natural ester dielectric fluid and novel insulation system for its use. IEEE/PES Transmission & Distribution Conference, New Orleans, April 12–16
24. Choi SH, Huh CS (2013) Analysis of the chemical and electrical characteristic of vegetable insulating oils exposed to accelerated aging. 2nd International Conference on Innovations in Electrical and Civil Engineering (ICIECE'2013) Dec. 17–18, Pattaya
25. Cigré working group A2.35: Technical brochure 436- Experiences in service with new insulating liquids
26. Lance R. Lewand, Report on the role of corrosive sulfur in transformers and transformer oil
27. Rapp KJ, Lemm AW, Orozco LA, McShane CP, Senior Member, IEEE, Corrosive Sulfur Phenomena Mitigation by Using Natural Ester Dielectric Fluids-Field Experience in Latin America
28. Forma MWE, Jungemann FA, Norris, Sonntag NO (1979) In: Swern D (ed) Bailey's industrial oil and fat Pmducts, vol l, 4th edn. Wiley, New York, pp 186–189
29. Noureddini H, Teoh BC and Davis L, Clement, (1992) Densities of vegetable oils and fatty acids, JAOCS, 69(12), 1184–1188.
30. Mike Rycroft (2014) features editor, EE Publishers, Energize-April 2014- p 37
31. Bertrand Y, Lauzevis P (2013) 22nd International Conference on Electricity Distribution Stockholm, 10–13 June 2013 [paper 0382]
32. OECD: Principles and strategies related to the testing of degradation of organic chemicals, http://www.oecd.org/chemicalsafety/testing/5598432.pdf.
33. Annual Book of ASTM Standards, Vol. 10.03 (1996) Electrical insulating liquids and gases; electrical protective equipment. American Society for Testing and Materials 1996
34. Lawate S, Unger R, Huang C (1999) Commercial Additives for Vegetable Lubricants. Lubricants World, May 1999, pp 43–45
35. Dunn RO, Shockley MW, Bagby MO (1996) Improving the low-temperature properties of alternative diesel fuels: vegetable oil-derived methyl esters. JAOCS 73(12):1719–1728
36. H2b, Ronny Fritsche, Uwe Rimmele, Frank Trautmann, and Dr. Michael Schäfer
37. Transformer fire protection, facilities instructions, standards, and techniques, Vol 3–32, Hydroelectric Research and Technical Services Group, Bureau of Reclamation, U.S. Dept. of Interior, Jan. 2005, Denver
38. 5–4 Transformers, Section 2.3.1.1.1 Separation Distance, Property Loss Prevention Data Sheets, Factory Mutual Global, May 2003, Norwood
39. Less-Flammable Liquid-Insulated Transformers, Article 450.23, National Electrical Code, NFPA 70, National Fire Protection Association, Quincy
40. Location and arrangement of power transformers and regulators, Section 152, National Electrical Safety Code, Institute of Electrical and Electronics Engineers, Piscataway
41. Classification of insulating liquids according to fire point and net calorific value, CEI/IEC 61100: 1992–05, International Electro technical Commission, p. 11, 2005, Geneva
42. Perret J, Paris M (1987) Silicone oils for transformers, E.D.F. Bulletin des études et recherches – Série B, Réseaux électriques, matériels électriques, No. 2 pp 5–13
43. Perrier C, Beroual A, Bessede JL (2006) Improvement of power transformers by using mixtures of mineral oil with synthetic esters. IEEE TDEI 13(3):556–564
44. Mc Shane CP, Rapp KJ, Corkran JL, Gauger GA, Luksich J (2001) Aging of Paper Insulation in Natural Ester Dielectric Fluid, EEE/PES Transmission & Distribution Conference, Oct. 28- Nov. 2, Atlanta
45. Mc Shane CP, Rapp KJ, Corkran JL, Luksich J (2003) Aging of Cotton/Kraft Blend Insulation Paper in Natural Ester Dielectric Fluid. Tech Con 2003 Asia Pacific, May 7–9, Sydney

46. Mc Shane CP, Corkran JL, Rapp KJ, Luksich J (2003) Aging of paper insulation retro filled with natural ester dielectric fluid. In: IEEE/DEIS Conference on Electrical Insulation and Dielectric Phenomena, Oct. 19–22, Albuquerque
47. Broekhuizen P, Theodor D, Le Blanch K (2003) Ullmer S (2003) Lubrication in inland and coastal water activities. A. A. Balkema Publishers, Tokyo
48. Honary LAT (2004) Biodegradable/Biobased Lubricants and Greases, Machinery Lubrication Magazine Issue No. 200109, Noria Corporation
49. Diaz E (2008) Microbial biodegradation: genomics and molecular biology, 1st edn. Caister Academic Press, Wymondham. www.horizonpress.com/biod
50. US Army Corps of Engineers (1999). US army Manual EM1110–2-1424 (chapter 8) Viewed from www.usace.army.mil/usace-docs/engmanuals/em1110-2-1424/c-8.pdf (Sept, 2008).
51. www.wiserenewables.com (2006). Wise Solutions–Renewable Lubricants–Biodegradability Primer.
52. Aluyor O, Ori-Jesu M (2008) The use of antioxidants in vegetable oils -a review. Afr J Biotechnol 7:4836–4842
53. Bioblend (2008). Biodegradable: What does that really mean? http://www.ignitemediahosting.com/.
54. Ferrari RA, Oliveira V, Scabio A (2004) Oxidative stability of biodiesel from soybean oil fatty acid ethyl esters. Sci Agric (Piracicaba, Braz) 62(3):291–295
55. Wolf WJ (1978) Soybeans and other oilseeds.In: Martin Grayson (executive editor) 1978 Kirk–Othmer encyclopedia of chemical technology, volume 21, 3rd edn. Wiley, New York
56. Gunther RC (1971) Lubrication. Bailey Brothers and Swinfen Ltd, Folkestone
57. Toshiyuki C (1999) Prediction of oxidative stability based in various chemical properties for refined vegetable oils. J Japan Oil Chem Soc 48(8):781–786
58. Gertz C, Klostermann S, Kochhar SP (2000) Testing and comparing oxidative stability of vegetable oils and fats at frying temperature. Eur J Lipid Sci Technol 102(8–9):543–541
59. Applewhite TH (1978) Fats and fatty oils.In: Martin Grayson (executive editor) 1978 Kirk–Othmer encyclopedia of chemical technology, volume 9, 3rd edn. Wiley, New York
60. Isbell TA, Abbot TP, Carlson KD (1999) Oxidative stability index of vegetable oils in binary mixtures with meadow foam oil. Ind Crop Prod 9(2):115–123
61. Jerzykiewicz M, Ćwieląg-Piasecka I, Jezierski A (2013) J Am Oil Chem Soc 90:803
62. Oyedeji FO, Oderinde RA (2006) Characterization of isopropanol extracted vegetable oils. J Appl Sci 6(11):2510–2513
63. Ullah J, Hamayoun M, Ahmad T, Ayub M, Zarafullah M (2003) Effect of light, natural and synthetic antioxidants on stability of edible oil and fats. Asian J Plant Sci 2(17–24):1192–1194
64. Emanuel NM, Lyaskovskaya YN (1967) The inhibition of fat oxidation processes. Pergamon Press Ltd, London
65. Dunn RO (2005) Effect of antioxidants on the oxidative stability of methyl Soyate (biodiesel). Fuel Proc Technol 86:1071–1085
66. Jung MY, Min DB (1990) Effects of α-γ- and β-Tocopherols on oxidative stability of soybean oil. J Food Sci 55(5):1464–1465
67. Ruger CW, Klinker EJ, Hammond EG (2002) Abilities of some antioxidants to stabilize soybean oil in industrial use conditions. J Am Oil Chem Soc 79(7):733–736
68. Senthil. Kumar S, (2016) Effect of antioxidants on critical properties of natural esters for liquid insulations, IEEE Trans Dielectr Electr Insul, 23(4): 2068–2078.
69. Khanahmadi M, Janfeshan K (2006) Study on Antioxidation property of Ferulago angulata plant. Asian J Plant Sci 5(3):521–526
70. Morteza-Semnani K, Saeedi M, Shahani S (2006) Antioxidant activity of the Methanolic extracts of some species of Phlomis and Stachys on sunflower oil. Afr J Biotechnol 5(24):2428–2432
71. Maduka HCC, Uhwache HM, Okoye ZSC (2003) Comparative study of the antioxidant effect of *Sacoglottis gabonensis* stem bark extract, a Nigerian alcoholic beverage additive and vita-

mins C and E on the Peroxidative deterioration of some common stored vegetable oils in Maiduguri. Pak J Biol Sci 6(3):202–207
72. Eastman Chemical Company (2001). Tenox Food-grade Antioxidants for Refined Vegetable Oils. www.eastman.com Eastman Chemical Company Kingsport Tennessee USA
73. Fennema OR (1985) Food chemistry, 2nd edn. Marcel Dekker, Inc., New York
74. Scott G (1965) Atmospheric oxidation and antioxidants. Elsevier, Amsterdam
75. Eastman Chemical Company, 2007
76. Ohio State University (2008) Antioxidants. http://class.fst.ohio-state.edu
77. Bennion M (1995) Introductory foods, 10th edn. Prentice-Hall Inc., Upper Saddle River
78. Martin D (2008) Evaluation of the dielectric capability of ester based oils for power transformers. Doctoral thesis, University of Manchester
79. Martin D, Wang ZD (2006) A comparative study of the impact of moisture on the dielectric capability of esters for large power transformers, Conf. on Electrical Insulation and Dielectric Phenomena, pp. 409412
80. Cargill Reference Data R2070- EnvirotempTM FR3TM Fluid–Dissolved Gas Guide, 2006
81. Frotscher R, Vukovic D, Jovalekic M., Tenbohlen S, Harthun J., Perrier C and M. Schafer (2012) Behaviour of Ester Liquids under Dielectric and Thermal Stress–From Laboratory Testi1ng to Practical use, CIGRE D1–105
82. Martin D, Lelekakis N, Davydov V (2010) Preliminary results for dissolved gas levels in a vegetable oil–filled power transformer. IEEE Electr Insul Mag 26(5):41–48
83. Wang Z, Wang X, Yi X, Li S, Hinshaw JV (2014) Gas generation in natural Ester and Mineral oil under partial discharge and sparking faults. IEEE Electr Insul Mag 29(5):62–70
84. Tenbohlen S, Vukovic D, Jovalekic M., Schäfer M, Harthun J (2009), Dielectric Performance and Dissolved Gas Analysis of Natural Esters for Application in Power Transformers, Performance of conventional and new materials for high voltage apparatus Cigre Sc D1 – Colloquium In Hungary Budapest
85. Zhongdong W, Xiao Y, Jinping H, Hinshaw JV, Noakhes J (2012) Fault gas generation in natural-ester fluid under localized thermal faults. IEEE Electr Insul Mag 28:45–56
86. Jovalekic M., Vukovic D, and Tenbohlen S (2011) Dissolved gas analysis of alternative dielectric fluids under thermal and electrical stress. IEEE International Conference on Dielectric Liquids (ICDL), pp 1–4, 26–30
87. Zafeiris K, Bertrand Y, Hipszki G, Kuffel M, Rehorek G, Seidel S (2011) Atanasova-Hoehlein I (2011) investigation on natural ester fluids concerning gassing and compatibility with transformer materials. International Symposium on High Voltage Engineering, Hannover
88. Imad UK, Zhongdong W, Cotton I, Northcote S (2007) Dissolved gas analysis of alternative fluids for power transformers. IEEE Electr Insul Mag 23:5–14
89. Wilhelm HM, Santos CC, Stocco GB (2013) Dissolved gas analysis (DGA) of natural ester insulating fluids with different chemical compositions. IEEE Trans Dielectr Electr Insul 21:1071–1078
90. Muhamad NA, Phung BT, Blackburn TR (2011) Dissolved gas analysis for common transformer faults in soy seed-based oil. IET Electr Power Appl 5:133–142
91. Perrier C, Marugan M, Beroual A (2012) DGA comparison between ester and mineral oils. IEEE Trans Dielectr Electr Insul 19:1609–1614
92. Dai J, Wang ZD (2008) A comparison of the impregnation of cellulose insulation by Ester and Mineral oil. IEEE Trans Dielectr Electr Insul 15(2)
93. Moser HP, Dahinden V (1987) Transformer board II, pp 23–25
94. Mc Shane CP, Luksich JRKJCJL, Gauger GA (2001) Aging of paper insulation in natural ester dielectric fluid. IEEE PES Transmission and Distribution Conference and Exposition 2:675–679
95. Mc Shane CP (2002) Vegetable-oil-based dielectric coolants. Industry Applications Magazine 8(3):8
96. Luksich J (2013) Loading guide A and B factors for envirotemp FR3 fluid and thermally upgraded kraft insulation. Report TP03-DR-009. Cooper Power System, p 3

97. Stockton D. P, Bland JR, Mc Clanahan T, Wilson J, Harris DL and McShane CP (2007) Natural Ester Transformer Fluids: Safety, reliability & environmental performance. Petroleum and Chemical Industry Technical Conference, p 7
98. Marulanda AR, Artigas MA, Gavidia A, Labarca F and Paz N (2008) Study of the vegetable oil as a substitute for mineral oils in distribution transformer. Transmission and Distribution Conference and Exposition: Latin America, 2008, IEEE/PES, p. 6
99. Mak J, Maciel L, Franchini LR, Vagner V (2006) Transformador de distribuição de maior vida útil e menor agressividade ambiental. IV Work Spot, CIGRE Brasil, p 8
100. Mc Shane CP, Martins MN, and Luksich J. (2006) Field experience with natural ester (vegetable oil) dielectric fluid-the first decade. IV Workspot-International Workshop on Power Transformers. CIGRE. Vol 18, pp 1–10

MIX
Papier aus verantwortungsvollen Quellen
Paper from responsible sources
FSC® C105338

If you have any concerns about our products,
you can contact us on
ProductSafety@springernature.com

In case Publisher is established outside the EU,
the EU authorized representative is:
Springer Nature Customer Service Center GmbH
Europaplatz 3, 69115 Heidelberg, Germany

Printed by Libri Plureos GmbH
in Hamburg, Germany